IMAGES OF WAR

THE BATTLE OF BRITAIN
LUFTWAFFE BLITZ

RARE PHOTOGRAPHS FROM WARTIME ARCHIVES

PHILIP KAPLAN

Pen & Sword
AVIATION

First published in Great Britain in 2013 by
PEN & SWORD AVIATION
an imprint of
Pen & Sword Books Ltd,
47 Church Street,
Barnsley, South Yorkshire.
S70 2AS

A CIP record for this book is available from
the British Library.

ISBN 978 1 78159 368 4

Printed and bound in England
By CPI Group (UK) Ltd, Croydon, CR0 4YY

Pen & Sword Books Ltd incorporates the imprints of
Pen & Sword Archaeology, Atlas, Aviation, Battleground,
Discovery, Family History, History, Maritime, Military,
Naval, Politics, Railways, Select, Social History, Transport,
True Crime, and Claymore Press, Frontline Books,
Leo Cooper, Praetorian Press, Remember When,
Seaforth Publishing and Wharncliffe.

For a complete list of Pen & Sword titles please contact
Pen & Sword Books Limited
47 Church Street, Barnsley,
South Yorkshire, S70 2AS, England
E-mail: enquiries@pen-and-sword.co.uk
Website: www.pen-and-sword.co.uk

Contents

The author is grateful to the following
for the use of their published and/or
unpublished material, or for their kind
assistance in the preparation of this
book: Malcolm Bates, Tony Bianchi,
K. Budzik, John Burgess, Winston
Churchill, Jack Currie, Al Deere, Bob
Doe, Hugh C.T. Dowding, Neville
Duke, Adolf Galland, Stephen Grey,
Roger Hall, Alex Henshaw, Eric
Holloway, Hargi and Neal Kaplan, Brian
Kingcome, Edith Kup, Eric Marsden,
Edward R. Murrow, Geoffrey Page,
Keith Park, Horst Petzschler, Andy
Saunders, William L. Shirer, and Peter
Townsend. Photographs from the
Bundesarchiv, the Imperial War
Museum, and the collection of the
author. Reasonable efforts have been
made to trace copyright holders to use
their material. The author apologizes
for any omissions. All reasonable efforts
will be made to correct such omissions
in future editions of this book.

The navigator and pilot of a Heinkel He 111 bomber attacking a target in England during the Battle of Britain.

In The Beginning

A lot of people over the years have tried to invade Britain, including the Romans in 55 BC. Later came the Vikings and the Danes. The Normans managed it successfully in 1066, but the attempt in 1588 by the King of Spain and his famous Spanish armada ended badly for him when his warships were defeated by Sir Francis Drake. Napoleon had a notion along those lines but was never able to assemble the necessary forces for such an undertaking.

Much of the "modern" history of Britain before 1940 revolved around her Royal Navy and how, for more than a thousand years, that navy had roamed the seas protecting the island nation. Her

left and right: British, French and Belgian soldiers rarely saw the Spitfire Mk Is and the Bf 109s of the Geman Luftwaffe high in the grey skies over Dunkirk in late May of 1940. It was the first encounter of the German pilots with the Mk I and they were shocked to find they were being both out-turned and out-climbed by their British opponents. Here, the main fire burning in the port, and wrecked cars and boats on the Dunkirk beach.

heroic admirals, Drake, Nelson and others had mostly triumphed over those who came by sea, engaging and shoving them back to their homelands, firmly defeated and pondering their failed efforts.

Then, in the summer of 1940, the German nation, fresh from her relatively easy conquest of Poland, Czechoslovakia, Austria, Belgium, Norway, the Netherlands, and France, confidently sent the Luftwaffe, her large and powerful air force, to clear the air over England and the English Channel in preparation for her own invasion attempt.

After the end of the First World War, the victorious Allies had punished their former enemy with a treaty executed at Versailles, a treaty put together over six months and signed on 28 June 1919 following the negotiations of that Paris Peace Conference. Among the most significant and controversial measures of the treaty were those compelling the Germans to accept responsibility for causing that war, to completely disarm, to make substantial territorial concessions, and to pay very heavy war reparations to the Allied countries that had opposed them in the war, a sum equivalent in 2013 to £284 billion. Noted economists of the time, including John Maynard Keynes, saw that last requirement as excessively harsh and unproductive. Those in leadership roles in Britain and the United States were inclined to be convinced by Keynes argument; the French, however, were not. A major intent of the treaty conference was the permanent pacification of Germany, along with the punishment dished out by the reparations element. These were key factors leading to the Second World War only twenty years later.

The negotiations leading to the Versailles Treaty had been organized to exclude the defeated Germany, Austria and Hungary, as well as Russia, which had negotiated a separate peace with Germany at the end of the war in 1918. Even the negotiators, the United Kingdom, the United States, France,

Italy, and Japan, admitted that the resulting treaty was harsh and punitive. The large delegations from each of these victor nations became so unwieldy that both Japan and Italy elected to exit the meetings, leaving only the representatives of Britain's David Lloyd George, France's Georges Clemenceau, and the U.S. President, Woodrow Wilson, to slowly hammer out what has come to be known as "the unhappy compromise."

At Versailles, the French expressed their position when Clemenceau said to Wilson, "America is far away, protected by the ocean. Not even Napoleon himself could touch England. You [the United States and Britain] are both sheltered. We are not." He referred to the fact that only France among them possessed a land border with Germany, and stated that the French wanted to take their border to the Rhine or, failing that, to create a buffer state in the Rhineland. The most the others would agree to was the de-militarization of the Rhineland, a mandate over the Saar, and the promises of Britain and the U.S. to support France in the event of a new aggression by Germany. As Keynes described the French position: "… it was the policy of France to set the clock back and undo what, since 1870, the progress of Germany had accomplished. By loss of territory and other measures her population was to be curtailed; but chiefly the economic system, upon which she depended for her new strength, the vast fabric built upon iron, coal, and transport must be destroyed. If France could seize, even in part, what Germany was compelled to drop, the inequality of strength between the two rivals for European hegemony might be remedied for generations."

Of the Big Three allies, France had suffered the greatest loss of life in the war, as well as immense damage and Clemenceau was utterly determined to achieve the payment of massive reparations by the Germans. To some extent he had the support in that demand of the British Prime Minister, Lloyd George, but the Briton also favoured the restoration of Germany to make her a strong trading partner, and thus he worried about the effect of the payment of those reparations on the British economy. Like the French, he was also concerned about the preservation of his empire. In the end he worked to increase the size of the reparations Germany had to pay, by couching that negotiation in terms of compensation for the great number of Allied war widows and orphans and the many men whose war injuries now prevented them from being able to work and earn a living.

Woodrow Wilson brought a liberal perspective to the talks. His agenda included free trade and self-determination for the Germans, rebuilding the German economy, and, of greatest importance to the United States, the creation of a powerful League of Nations to keep the future peace. Wilson disagreed with the British and French harsh position on Germany, but was outvoted by them.

Among the key clauses of the treaty was that dealing with arms limitation. "In order to render possible the initiation of a general limitation of the armaments of all nations, Germany undertakes strictly to observe the military, naval, and air clauses which follow.

"German armed forces will number no more than 100,000 troops, and conscription will be abolished.

"Enlisted men will be retained for at least twelve years; officers to be retained for at least twenty-five years.

"German naval forces will be limited to 15,000 men, six battleships (no more than 10,000 tons displacement each, six cruisers (no more than 6,000 tons displacement each), twelve destroyers (no more than 800 tons displacement each), and twelve torpedo boats (no more than 200 tons displacement each). No submarines are to be included.

"The import and export of weapons is prohibited.

Reichsmarschal Hermann Goering in France prior to the start of the Battle of Britain; below left: Generaloberst Hans Jeschonnek; below right: Generalfeldmarschal Albert Kesselring.

"Poison gas, armed aircraft, tanks and armoured cars are prohibited.

"Blockades on ships are prohibited.

"Restrictions on the manufacture of machine-guns (e.g. the Maxim machine-gun) and rifles (e.g. the Gewehr 98 rifle).

Under reparations in the main article of the treaty, Article 231 laid the blame on Germany for the war and established the reparations that she must pay the Allies, the aforementioned equivalent sum in 2013 of £284 billion. Those reparations would be paid partially in currency and partially in the form of steel, coal, agricultural products, and intellectual property. The Allies agreed to minimize the amount of currency in the reparations in order to hold down the resulting effects of hyper-inflation which would in turn reduce the reparation income to the French and British.

One of the British representatives at the Versailles Treaty conference, James Ramsey MacDonald, who would later become the first Labour Prime Minister of the United Kingdom, remarked after the conference that France's policy in the proceedings had been greedy and vindictive. A widely held view

Heinkel He 111s in France in 1940; right: Bombing the Thame-shaven oil storage tank farm.

in France, on the other hand, as expressed by Major General Ferdinand Foch was that, if anything, the restrictions of the treaty on the Germans were too lenient: "This is not Peace. It is an Armistice for twenty years."

In late April 1919, a German delegation headed by the Foreign Minister Ulrich Graf von Brockdorff-Rantzau went to Versailles where they were presented with the conditions that had been established by the victorious Allies. On reading them, his response to Lloyd George, Clemenceau and Wilson was: "We know the full brunt of hate that confronts us here. You demand from us to confess we were the only guilty party of war; such a confession in my mouth would be a lie." The German government protested the demands it saw as unfair and "a violation of honour." Prominent Germans across the political spectrum attacked the treaty—especially the provision blaming Germany for starting the war.

In Germany itself, blame was broadly levelled at those perceived as opposing German nationalism, instigating unrest, or profiteering. The failure of the final German campaign of the war, the Spring Offensive, was blamed on industrial strike action in the German armaments industry during the offen-

sive, which limited the supply of materiél to the front-line troops.

With the early payment of the reparations by the German government, the German economy was gripped by hyper-inflation and, as conditions in Germany worsened, a range of violations or treaty provision avoidances began. These included the dissolution of the German General Staff in 1919, a cover for the creation of the German *Truppenamt*, a military organization established to rewrite the German doctrinal and training materials, incorporating the lessons of the First World War. In the 1930s the German government responded to what it declared to be the Allies' own violation of the Versailles Treaty in failing to initiate required military limitations on themselves. The Germans announced that they would cease adhering to the treaty's military limitations. Then came the accession to power in Germany by Adolf Hitler and the National Socialists, who violated the treaty by instituting compulsory military conscription in March 1935. Two months later, the British withdrew from the treaty and signed a new Anglo-German Naval Agreement. In March 1936, another German violation of the treaty occurred when they reoccupied the demilitarized zone in the Rhineland. In March 1938, they annexed Austria. In September, with the cooperation and approval of Britain, France and Italy, the Germans again violated the treaty when they annexed the Sudetenland from Czechoslovakia . . . in March 1939, their annexation of the remainder of Czechoslovakia, and on 1 September the German army and air force launched an invasion of Poland, which began the Second World War.

above: Generalfeldmarschal Erhard Milch who was given the title Air Inspector General and placed in charge of aircraft production for the Luftwaffe; centre: Luftwaffe armourers preparing to load a bomb on a French airfield.

The pilot of a Messerschmitt Bf 110 is assisted by his ground crew before a raid on England in August 1940.

Joachim von Ribbontrop greeting Btitish Prime Minister Neville Chamberlain in 1938; centre: Factory assembly of Heinkel He 111 bombers in 1939; far right: Production of Dornier Do 17 bombers.

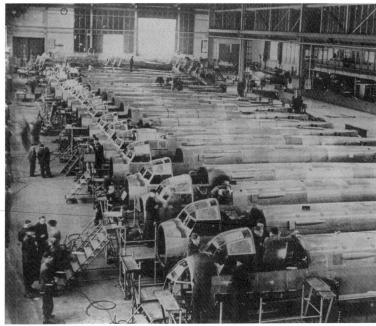

Preparation

In the company of his boss, the new Nazi leader Adolf Hitler, Hermann Goering, who would in several years command a new German Air Force in the Battle of Britain, addressed a jubilant crowd in Berlin's Wilhelmstrasse: "My German comrades, January 30th, 1933, will enter German history as the day on which the nation, after fourteen years of torture, need, deformation, and shame, has found its way back to itself . . . the future will bring everything for which the Feuhrer and his movement . . . have fought . . . in spite of all reverses and disappointments . . ."

From that moment in the German capitol, Goering devoted himself to the goal of developing for the nation a new and powerful air force capable of being a decisive factor in a European war. Goering had long been a disciple of the philosophy of Italian General Guilio Douhet and his theories of total air warfare. Douhet had been a key proponent of strategic bombing in aerial warfare and was among the earliest advocates of the creation of a separate air arm commanded by airmen rather than ground commanders. After being imprisoned in Italy for one year for having criticized the Italian military leaders in the First World War, he was released, exonerated and promoted to the rank of general in 1921. In that same year he completed and published his highly influential treatise on strategic bombing, The Command of the Air, a work that greatly impressed Goering.

In fact, it took much more than Goering alone to design and build the Luftwaffe. He had the guidance of General Hans von Seekt, who, as a member of the German peace delegation after the First World War had early exposure to the vengeance dished out by the Allies in the Versailles Treaty conditions. Seekt was determined that Germany would not adhere to those conditions and his assistance, coupled with the thoughts and ideas of Douhet, and the clandestine decade-long training of German military pilots and crew at Lipetsk in the Soviet Union, paved the way for the future German

A Junkers Ju 87 Stuka dive-bomber assembly line in 1939.

founding director of the airline, Deutsche Luft Hansa. In 1933, Erhard Milch became State Secretary of the *Reichsluftfahrtministerium* [Reich Aviation Ministry], in which capacity he worked directly for Goering. No less a figure in German aviation than Dr Ernst Heinkel, the aircraft designer and manufacturer, said of Milch: "…he possessed great capabilities and unbounded personal ambition and a ruthless energy. Milch was disinclined to leave his high position with the airline when summoned by Goering, but agreed to do so when interviewed by Hitler." Milch: "Hitler took a man's soul from him. He was amazingly quick at grasping technical details. He knew much more than Goering."

The Lipetsk training, and the access to technical personnel from Luft Hansa, helped form a solid basis for the air force that Hitler and Goering wanted, but what was still missing was a core of professional airmen suited for high command in the new Luftwaffe. To fill that gap, Seekt brought in and groomed a handful of particular specialists that included Albert Kesselring to head Supply and Organization, Hans-Juergen Stumpff to run Personnel, Walther Wever as Chief of Operations, Hugo Sperrle, and Hans Jeschonnek. Of these, only Jeschonnek had actually piloted an aircraft; the rest, though, were eager to learn.

In the spring of 1933, the Nazi government created the new Air Ministry, installing Goering as Air Minister with Milch as Secretary of State for Air. When Goering assembled seventy of his most promising young pilots in a meeting at the Air Ministry, he told them "the time has come to throw off the chains of Versailles. Now that the secret training in Russia is over, Mussolini will help out for the time being with the training of our fighter pilots. But in order to avoid complications you will go to Italy under the strictest secrecy." One among this group of emerging German airmen was Adolf Galland, who would become one of Germany's greatest fighter pilots and *General der Jagdflieger*, Commander of the German Fighter Force. In 705 combat sorties, Galland was credited with 104 aerial victories on the Western Front and in the Defense of the Reich. After the war he met and befriended a number of his former enemies, among them the Battle of Britain aces Robert Stanford Tuck and Douglas Bader.

When two dozen members of both houses of the British Parliament flew their own private planes to Düsseldorf during Whitsuntide of 1933, they were met by dozens of Nazi brownshirts, carefully shepherded and shown only the vaguest indication of Germany's achievements in aviation to that point, and certainly no hint of the developments soon to come like the Me 109 fighter of Willy Messerschmitt. All was corgial at the Aero-Club von Deutschland dinner given for them that evening by Hermann Goering.

In a little-known irony, the aero engine makers of both Germany and Britain had a part in advancing the military aircraft development of their future enemy. British Napier engines powered the Fokker D.13, and the Rolls-Royce Kestrel was the powerplant of the RAF Fury fighter, and used by Arado in the Ar. 67. Early examples of the Junkers Ju-87 Stuka dive-bomber and the Messerschmitt Me 109 fighter both flew with Rolls-Royce engines; and the Merlin first powered a German aircraft acquired from Heinkel, the Heinkel He 70 Blitz. Thus it was that the German plane maker unintentionally aided Rolls-Royce in the development of the engine that would ultimately power the Spitfire, Hurricane, Lancaster, Mosquito, Mustang, and other aircraft that would play a major part in Germany's defeat in the Second World War.

In Britain, meanwhile, technical development in the Royal Air Force had, since 1930, been in the capable hands of Air Vice Marshal Sir Hugh Dowding, in charge of Research and Development. Sir Cyril Newell had charge of Supply and Organization. Dowding put Captain F.W. Hill in charge of the Armament Experimental Establishment at Martlesham, of whom Dowding said: "Just the sort of chap who would get a new idea five years before anybody else." In a meeting of armament specialists on 19 July 1934, Hill told them that the new fighter required by the RAF would need eight machine-guns firing at a thousand rounds a minute to destroy a bomber in two seconds. This pronouncement led R.J. Mitchell, the designer of the Schneider Trophy-winning racer, and Sydney Camm, chief designer of the Hawker Company, to scrap the designs they had been working on for a new four-gun monoplane fighter, and concentrate on a new Air Ministry specification for an eight-gun fighter with a closed cockpit and retractable undercarriage. In a separate announcement of 19 July, the British government said that it planned to expand the RAF by forty-one squadrons. Many of these new outfits would be equipped with Mitchell's Spitfires and Camm's Hurricanes.

Back on the continent, the German pretense of respecting and obeying the provisions of the Versailles Treaty continued, an example being the naming of the Luftwaffe's first bomber squadron, the *Vershuchsanstalt für Schaedlingsbekaempfung*—literally the Agricultural Pest Control Unit.

In July Winston Churchill stood in the Commons to express his thoughts on the vulnerability of Britain to attack, "... with our enormous Metropolis here, the greatest target in the world, a kind of tremendous fat cow ... tied up to attract the beasts of prey." Churchill knew from "private sources" that Germany already had put together a significant military air force, and that that air force would

quite soon be stronger than the Royal Air Force. As usual in those days, his words were greeted largely with derision. The policy of the British government had focused on the expansion of an effective home-based bomber force for the RAF, a force they hoped and expected would pose sufficient threat to the Germans to make them reconsider before acting on any plans to attack Britain or France. Prime Minister Chamberlain, both in that role and his former one of Chancellor of the Exchequer, had refused to fund a broader, more balanced defence strategy which would have included the formation of an Expeditionary Force, choosing instead to depend almost entirely on the deterrence of the bomber threat to stave off Hitler.

The weakness of Britain's military policy persisted from 1934 through the 1938 when the Munich Agreement in which Britain and France permitted the Nazis to annexe areas along Czechoslovakia's borders which were largely inhabited by ethnic Germans, areas the German government now referred to as Sudetenland. The Czechs had not been invited to the Munich conference and felt betrayed by the British and French. Ultimately, the British were forced to assemble an Expeditionary Force to be sent to Europe with the outbreak of the Second World War. The British bomber-based military policy by the time of the Munich Agreement, had failed to produce the strong bomber force the government had wanted.

By the first week of September 1939, when the Germans had invaded Poland, and the British and French had declared war on Germany, the vaunted British bomber force was utterly deficient in front-

left: A formation of Heinkel He 111H bombers en route to an English target in the summer of 1940. The He 111 was the principal bomber of the German Air Force in the early years of the Second World War; below: Some of the Chain Home High tower masts along the English sea coast which made up part of Britain's very effective Chain Home radar defence system that was completed in 1940.

Pick As ist Trumpf

above: Engine maintenance on a Dornier Do 17 medium bomber at a French airfield; below left: Masts of the Chain Home Low radar system above chalk cliffs of the English Channel coast; Quality artwork applied to a Luftwaffe bomber engine nacelle; right: The application of enemy shipping kill markings to the tail of a Luftwaffe bomber; far right: An example of nose art on a Messerschmitt Bf 110 fighter.

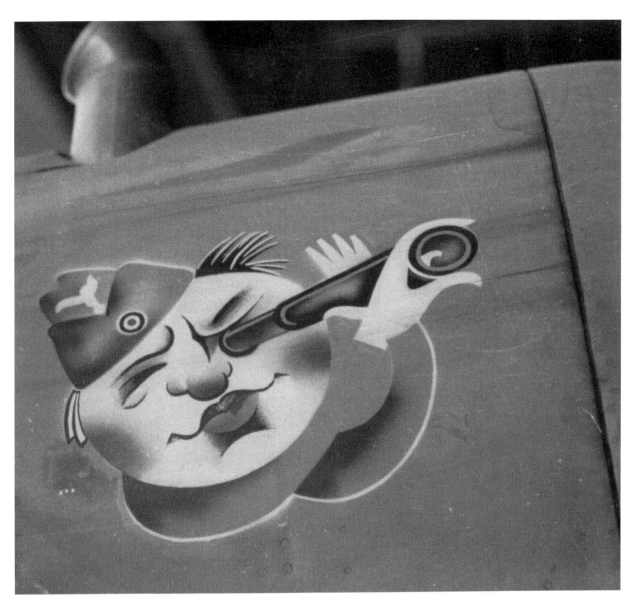

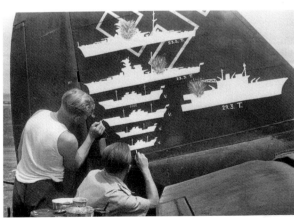

line strength, with mostly obsolete aircraft and crews incapable of reaching the well-defended German targets in daylight, or locating them at night as they lacked the navigational aids that would not be available to them until much later in the war. Britain's situation was made worse for neglecting to properly provide her late-blooming expeditionary force with the substantial air support it would require. In the plus column, though, was the basic development and deployment of a solid air defence system for Britain. A prime mover in that capability was Sir Kingsley Wood who, in May 1938, replaced Lord Swinton as Secretary of State for Air. Wood was firm in his insistence that the fundamental emphasis of the British aircraft industry be the equipment needs of the RAF fighter force. Swinton's main contribution to the air defence of Britain was his part in the development of the early-warning radar capability, initially known as radio direction finding, which enabled the interception of rapidly incoming enemy aircraft. The men who pioneered radar for the British defence system were Robert Watson-Watt, Henry Tizard, and Air Chief Marshal Sir Hugh Dowding, the Air Member for Research and Development, who a year later would take control of British air defences.

Goering and the Germans shared the view of the British Air Staff about the decisive importance of strategic bombing, but they shifted somewhat from that position following their experience in the Spanish Civil War, deciding that the most important role for the Luftwaffe was to spearhead the way for an advancing German army. To that end, they effectively shaped the German air force, to the extent, however, that the emphasis on tactical support began to far outweigh the ability to meet the strategic part of their remit. That imbalance of emphasis would in the end cost them the Battle of Britain. They had ridden the crest of relatively easy successes in the Scandinavian, French, and Low Countries campaigns, working closely with German ground commanders to smoothly advance the tactical situations there. But when they were required to do the business of preparing for the German invasion of Britain, they were unable to meet the need and forfeited their advantage of numerical superiority.

If Air Chief Marshal Dowding was perceived in any way as flawed when he was passed over for the job of Chief of the Air Staff a few years before the outbreak of World War Two, it would have been because he didn't subscribe to the prevailing Air Ministry philosophy about the invincibility of the bomber and was thought to be too defensively orientated. It was, of course, fortunate for Britain that Dowding instead became the commander of RAF Fighter Command during the Battle of Britain, virtually perfect type-casting for the air defence of the nation in its greatest ever hour of need.

The competent air leaders of the German Air Force, while able and efficient, were no match for Dowding and his commander of Eleven Group, Fighter Command, in the southeast of England, New Zealander Air Vice Marshal Keith Park. A superb tactician and manager of his resources, Park wielded his fighter squadrons with subtlety and restraint in the Battle of Britain, nearly always pouncing on the enemy air fleets at the best possible moment to punish them decisively and in the extreme. He operated in that way throughout the Battle, despite the persistent criticism of many in the government and the RAF who wanted him to prematurely commit the entirety of his squadrons. In Park, Dowding had a gifted area commander, ideally positioned to defend and strike at the enemy air force in its attacks on the prize target, London.

The way to an efficient, reliable early warning system for Britain in the 1930s was not a smooth and easy one. Air exercises by the RAF in the summer of 1934 had shown up the many weaknesses in the early warning system of that time. That system was the Adcock-Chandler method of radio direction finding, a promising concept meant to let defending fighters be 'fixed', 'plotted', and precisely

A German bomb aimer's view from the nose of his Heinkel bomber during the Battle of Britain.

positioned by ground controllers. But it relied upon frequent transmissions from the fighters; a contribution that could not be expected from hostile attacking bombers. During his researches into trying to discover clues to developing a greatly improved detection process, A.P. Rowe of the Air Ministry Directorate of Scientific Research, wrote to his boss, H.E. Wimperis: "Unless science finds some new method of assisting air defence, any war within ten years will be lost."

In June 1932, a British Post Office report, No 233, had noted that aircraft interfered with radio signals and re-radiated them. It had been written by Robert Watson-Watt. In 1922, Watson-Watt had received one of two American cathode-ray tubes at Farnborough. Wimperis now asked Watson-Watt to urgently look at the matter of aircraft detection and even at the possibility of developing "death rays."

Henry Tizard, meanwhile, had been given the chair of an Air Ministry scientific committee charged with a survey of air defence. When the committee gathered for the first time, in January 1935, Watson-Watt told them that death rays were out of the question, but he had made some progress on the issue of aircraft detection. He stated that a new understanding of the height of the ionosphere, coupled with the report that aircraft reflected radio signals, led him to conclude that a cathode-ray tube could be made to show the distance and height of an aircraft target. His report impressed the Tizard committee sufficiently to cause them to grant him £10,000 to develop his idea.

26 February. Late in that afternoon, in a field near the high Daventry transmitter masts, Dr Watson-Watt, A.P. Rowe, and two assistants huddled around a luminous screen of a crude, early type of television set in a trailer. They heard the sound of an approaching Heyford bomber, and as they watched the screen and heard the plane, a green spot in the middle of the screen grew larger and then smaller as the sound of the Heyford diminished. Watson-Watt had created the basis for the device that would be essential to the defeat of the Germans in the Battle of Britain.

10 March. Hermann Goering told the world, via the London Daily Mail, that "The objective was not the creation of an offensive weapon threatening other nations, but rather a . . . military aviation strong enough to repulse attacks on Germany." Demanding equality in the air, he announced "A new German air force has stepped onto the scene of world politics."

16 March. The Nazi Propaganda Minister Joseph Goebbels disclosed the passage of a new German law "for the creation of national defense forces" and said that Hitler had discarded the portions of the Versailles Treaty dealing with German armed forces. The following day, the American correspondent and author William L. Shirer wrote that it was "a day of rejoicing in Germany. The shackles of Versailles, symbol of Germany's defeat and humiliation, had been torn off." One week later, in a Berlin meeting between Hitler and William Strang of the British Foreign Office, Strang asked the German leader, "What is the strength of the German air force?" Hitler replied "We have reached parity with Britain." The Luftwaffe was 'official' and out in the open.

The autumn of 1935 saw the maiden flights of three aircraft types that would have a major impact on the Battle of Britain and the German bombing Blitz on London. On 6 November, K5083, the prototype of the Rolls-Royce Merlin-engined Hawker Hurricane fighter flew for the first time. The Hurricane would account for about two-thirds of all the German aircraft shot down during the Battle. She is remembered as the somewhat less lovely sister of the lovlier, more glamourous Vickers-Supermarine Spitfire, the aeroplane still revered by airmen the world over as either the one machine they have most enjoyed flying, or the one they would most like to fly. But the Hurricane, while not quite

as fast or as fabulously manoeuvrable as the Spitfire, was a much better and steadier gun platform, which, after all, is what a fighter plane is meant to be. And, very importantly, she was partially fabric-covered, making the repair of bullet holes a much simpler and less specialized procedure than on the all-metal fuselage and wings of the Spit, which normally had to be sent to a special Spitfire repair facility.

In that same season, the maiden flights of two vitally important Luftwaffe aircraft took place—the principal German fighter of the Battle and the Blitz, the Messerschmitt Me 109, and the screaming, vertical-diving Junkers Ju-87 Stuka. Because the Jumo engines meant to power both of these aircraft were not ready for installation and flight testing in time for the scheduled first flights, the manufacturers instead mounted Rolls-Royce Kestrel powerplants in them for these occasions. Across the English Channel, meanwhile, Rolls-Royce was utilizing the Heinkel 70 it had purchased, as a flying test-bed for the new Merlin aero engine.

above: In the cockpit of the Junkers Ju 87 Stuka dive-bomber. Operated by a pilot and a rear gunner, the Stuka was an impressive symbol of German air power early in WW2, but suffered from lack of manoevrability and speed, dooming it to failure in the Battle of Britain period. It is remembered for its wailing "Jericho Trumpet" siren.

It Starts

Probably no one in the world knew more about the operation of an air force fighter command than Hugh Dowding. Air defence and all the other aspects of the Fighter Command, including operations, training, research and development, and the myriad problems associated with RAF Fighter Command, had been his area of expertise for nine years when he moved to its new headquarters north of London at Bentley Priory, Stanmore in July 1936. By then he had championed and powerfully supported the case for and development of radar and the eight-gunned fighter. Dowding was an air defence expert. He had long disagreed with the widely held air force philosophy that 'attack is better than defence' and the primary emphasis of the RAF should always be on the bomber force, as advocated by Marshal of the Royal Air Force, and one of the original organizers of that institution, Hugh 'Boom' Trenchard. Dowding: "Trenchard seemed to have forgotten that 'security of the base' is an essential prerequisite. Since I was a child, I have never accepted ideas purely because they were orthodox, and consequently I frequently found myself in opposition to generally accepted views."

Dowding was a perfect choice to run Fighter Command. He immediately got to grips with the staffing and equipment requirements, and the various other needs to get the operation up and running. Of the many deficiencies he found when he took charge: "The most crying need was for Operations Rooms at all Commands and Stations, with tables on which courses of all aircraft, hostile and friendly, could be tracked ... there was absolutely no establishment for the manning of any Operations room. In the silly exercises which were sometimes held in the long evenings of summer the [Unit] Commander himself acted as Controller and his staff had to man the Operations Room ... these duties would have to be carried out twenty-four hours a day.

"There had not been any attempt ... to represent our own bombers leaving and returning to the country. Everything on the table was assumed to be hostile." Other problems included the vital Observer Corps, which was comprised of volunteers who were trained in the evenings after their workday and for whom there was no specific mobilization scheme and no authorization for pay. Another factor was the frequent refusal by the Air Staff to provide 'friendly bombers' for the training exercises. Among Dowding's greatest needs was that of all-weather runways. He knew that one of his key airfields, Kenley, part of the Kenley-Croydon-Biggin Hill sector which would have primary responsibility for the protection of London in the Battle of Britain and the Blitz period, would be out of use for three months during the coming winter. Again, he faced a recalcitrant Air Staff which had brought in a specialist army officer who devised an airfield camouflage scheme for Kenley. The Air Staff objected to the request for permanent all-weather runways on the basis that they would spoil the camouflage.

Dowding was constantly guided by his unwavering determination to make 'the base'—Britain—safe and secure. Better than anyone, he knew the scope of the threat facing the country in the form of the German air force, and all that had to be done to meet that threat, and how time was running against him. His needs were measured in men and aircraft, buildings and station facilities, and telecommunications equipment including the new, largely unproven radar. By the autumn of 1937, test results and RAF exercises had been encouraging for the continuing development of British radar, but work on the system down in Suffolk, Essex, and Kent, which was now known as CH, Chain Home, while well under way, was a long way from completion and real operation. To function properly the system

left: Luftwaffe Dornier Do 17 bombers leaving French airspace on their way to their target in England in the summer of 1940; below: Citizens of Portsmouth, England, who have volunteered to fill sandbags at Southsea Beach in 1939 to protect buildings in their town from the effects of enemy bombing.

would require twenty-one stations and about two years' work to finish. Dr Watson-Watt was having to endure the same sort of governmental red tape and delays that Dowding had been experiencing. Desperately concerned and frustrated, Watson-Watt went to see Winston Churchill about the problem. A few days later the log-jam was broken and the twenty-one station project approved and fully funded by the Treasury. At last the urgency behind the project seemed to have been realized and accepted at the various levels of authority and the work on Chain Home went forward at a far more serious pace.

When General Erhard Milch was invited to visit RAF Croydon in October 1937, he met several of the British air force figures of the time, but was especially impressed by the ordinary squadron pilots and cadets he encountered during his brief visit. "I could find no difference between your boys and ours—the same quality, the same spirit. Both the German and British fighter pilots had that same lightheartedness and bantering talk. They were really brothers by nature." But another thought came to him in that visit. "England had the training resources of her Empire and I wondered what would happen if war came. In the Luftwaffe we had no experienced leaders."

above: Three members of a Heinkel bomber crew during a mission to bomb a Royal Air Force Fighter Command airfield in England in the summer of 1940; right: A German bomb aimer at his work station.

top: RAF fighter pilots trying to relax on their airfield while awaiting the call to 'scramble' and intercept enemy aircraft on an incoming raid.

Milch's concern was increased when, on his return to Germany to report to the Luftwaffe chief, Goering showed little interest. Goering had for some time been resentful of the growing prestige and enhanced reputation of Milch, who was becoming ever more linked with himself as the guiding forces of the new German air force. Months passed with Goering unwilling to meet with Milch. In that period Goering shifted control of the Air Staff, personnel, and the technical departments from Milch and engineered a series of key personnel changes designed to hamstring his number two, after which Milch informed Goering "It's time I went. Apparently I have not done my job properly and I should like to go back to Lufthansa," to which Goering replied, "On the contrary, you have done your job too well. Everyone thinks you are the head of the Luftwaffe. I'll not let you resign, and don't go and pretend you are sick. If you want to commit suicide, go ahead. Otherwise you'll stay where you are." Hitler then called and asked Milch to come see him and tell him about his trip to England.

In August 1938, Goering hosted a visit to Berlin by General Joseph Vuillemin, Chief of Staff of the French Armée de l'Air. The general was shown the base facilities that his host wanted him to see, and spent a day with the Luftwaffe's elite *Jagdgeschwader Richthofen* squadron, Goering's old unit in the First World War. Vuillemin was duly impressed by what he saw and the personnel he met. He then visited the Augsburg and Leipzig Messerschmitt factories to look at the assembly lines of the Me 109 and Me 110 front-line fighters and made a depressing mental comparison of what he was seeing, with the current French air force. He was treated to an air display at Barth where the latest Luftwaffe bombers and dive-bombers performed with remarkable skill and precision. The visit left the French air chief glum and wholly convinced that if war broke out between Germany and France, the Germans would defeat the French air force in no more than two weeks.

By late August, it was clear to those in Luftwaffe headquarters that the air force would be severely stretched in support of the army when it moved against Czechoslovakia, as the Nazi master plan intended. The prevailing view was that, should the British intervene, the Luftwaffe would be incapable of establishing an effective blockade and might well also be required to attack France then. There was concern that the Luftwaffe bombers, operating from bases in Germany, lacked both range and bombload capacity to effectively attack targets in England. The prime targets, they believed, were the key warships of the Royal Navy which, they expected, would be moved well out of reach of the German bombers. Thus, the most worthwhile targets left to the Luftwaffe bombers would likely be the fighter airfields of the Royal Air Force and the aircraft factories in the London area. It was a sobering realization for the Luftwaffe planners that they were, in fact, not really capable of the sort of strategic bombing attacks on England that might be required of them. The actual limit of their capability then seemed to be the tactical support of the German army.

Based largely on General Vuillemin's report on his Luftwaffe tour, the French now were genuinely alarmed about their prospects should they have to face the full might and fury of the German air force. Across the Channel, the British Air Minister Kingsley Wood minced no words when he described the inadequacy of Britain's air defences to that point. He told government ministers and the opposition that the German bomber force was truly threatening and, if an attack came, the British should expect to suffer upwards of 500,000 casualties in the first three weeks. Wood and the government officials were actually misinformed about the relative strength of the German air force at that moment. While Prime Minister Chamberlain believed—or may have actually known—that Wood's information was exaggerated, he did not say so to his nervous Cabinet when they met to authorize him to negotiate

left: A German guard in front of the former RAF headquarters building in the occupied Channel islands in summer 1940; below: Refuelling a Dornier Do 17 bomber on a French airfield during the Battle of Britain.

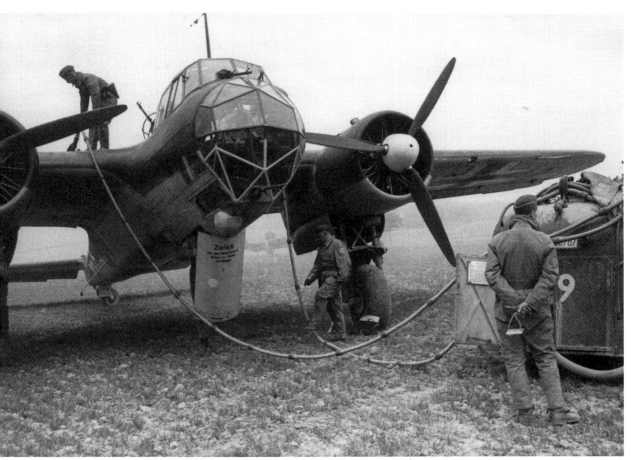

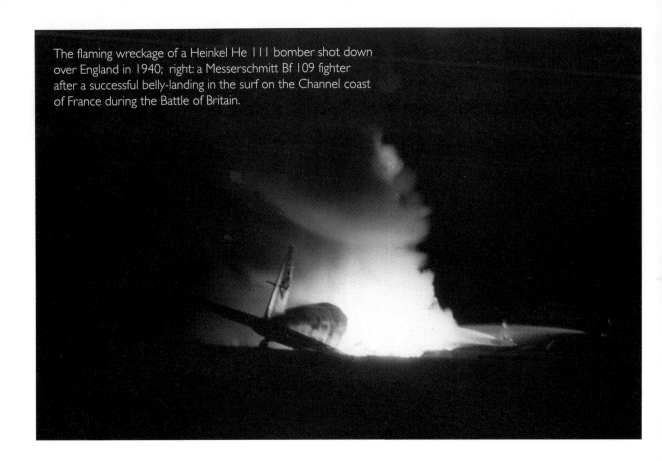

The flaming wreckage of a Heinkel He 111 bomber shot down over England in 1940; right: a Messerschmitt Bf 109 fighter after a successful belly-landing in the surf on the Channel coast of France during the Battle of Britain.

with Hitler and, in the process, sacrifice Czechoslovakia. Regardless of the true bomber strength of the German air force, its intimidation value, especially in light of the far weaker French and British air forces then, was enormous and it greatly influenced the positions taken by both France and Britain in the September Munich Conference. With that infamous appeasement agreement on 30 September, Neville Chamberlain returned to Heston aerodrome, London, that afternoon, where he waved the paper with Hitler's signature, and proclaimed "peace in our time." There were palpable sighs of relief on fighter stations around southern England among the young RAF pilots who knew full well that, had they been called upon to go up against the Luftwaffe, they had only ninety modern Hurricanes out of the 750 Fighter Command aircraft then available for action. All the rest were obsolescent biplanes.

Spurred by the military planning mistakes, misjudgements, and ultra-conservative attitudes of the British government of recent years, the Air Minister and his staff went to work immediately after the Munich Conference, to re-equip RAF Fighter Command on a highest-priority basis. Group Captain Peter Townsend: "By mid-December we [43 Squadron at Tangmere] had our full initial equipment of sixteen aircraft. The Fury had been a delightful plaything; the Hurricane was a thoroughly warlike machine, rock solid as a platform for its eight Browning machine guns, highly manoeuverable despite its large proportions, and with an excellent view from the cockpit. The Hurricane lacked the speed and glamour of the Spitfire and was slower than the Me 109, whose pilots were to develop contempt for it and a snobbish preference for being shot down by Spitfires. But figures were to prove that during the Battle of Britain, machine for machine, the Hurricane would acquit itself every bit as well as the

Spitfire and in the aggregate (there were more than three Hurricanes to two Spitfires) do greater execution among the Luftwaffe."

The Germans reoccupied the demilitarized zone in the Rhineland in March 1936. Two years later they annexed Austria and, in September 1938, they annexed the Sudetenland of Czechoslovakia, followed in March 1939 when they took the remainder of Czechoslovakia. On 1 September they invaded Poland, which action caused Britain and France to declare war on Germany on the 3rd.

Before the war began, the British Chief of the Air Staff, Cyril Newall, and his team had estimated the minimum requirement for Fighter Command was forty-six squadrons for general defence and an additional six squadrons to protect coastal convoys and the Scapa Flow naval anchorage in the Orkney Islands, Scotland. But on 3 September Fighter Command had just thirty-five squadrons, of which four had flown their Hurricanes to France, with a further six squadrons on stand-by for deployment to that war front. Now Dowding was faced with just twenty-five squadrons—less than half the number he required for the defence of Britain. He was desperate for the units he lacked and, when he requested an additional twelve squadrons as soon as possible, and the request was not granted, he was amazed when, in a meeting of 17 October, the Air Staff chief told the group that eight new squadrons would be ready for Fighter Command within the next two weeks and, incredibly, within the following two weeks a further ten squadrons would be formed. It was through his action that day that a narrow vic-

tory was ensured for the RAF in the Battle of Britain the following summer.

By November 1939, Hitler laid out his intentions. "My decision is unchangeable. I shall attack England at the most favorable and quickest moment. Breach of the neutrality of Belgium and France is meaningless . . . Victory or defeat! The question is . . . who is going to dominate Europe in the future . . . I have led the German people to a great height, even if the world does hate us now. I am setting this work on a gamble . . . I shall shrink from nothing and shall destroy everyone who is opposed to me . . ." A new Hitler directive of 29 November stated "In our fight against the Western Powers, England has shown herself to be the animator of the fighting spirit of the enemy and the leading power. The defeat of England is essential to victory." To achieve that aim it would be essential for the Luftwaffe to destroy the Royal Air Force.

In January 1940, Peter Townsend's 43 Squadron had relocated from Tangmere on the Sussex coast, to Acklington in Northumberland. Lacking the luxury of a resident snowplow, it fell to the squadron's

left: Members of the Observer Corps in their coast site in Kent; right: Junkers Ju 87 dive-bombers over southern England,

Hurricane pilots to improvise a way to clear a runway so they could fly. They pulled a door off a dispersal hut and had six men dressed in thick woolies and Balaclavas sit on the door while they towed the whole rig behind a tractor up and down the runway until the thick layer of snow had been thoroughly flattened into a hard surface.

No 43 Squadron had been sent north to Acklington on 18 November to fly patrols over the British merchant coastal convoys, frequently in the most foul weather. Never had they been more conscious of that Rolls-Royce Merlin engine ticking over in their Hurricanes as they flew above the icy sea. If that great engine should quit they would be down in the freezing water, floating in a Mae West life jacket, but without a dinghy, for they had not yet been provided. Their chance of survival was practically nil.

The crew of a Heinkel He 111 bomber of 2nd Gruppe, KG26 had been sent from their Westerland base on the Isle of Sylt to Schleswig in northern Germany near the Danish border, following a report of a southbound British convoy out of Sweden on 2 February. At Schleswig they were told to

left: Messerschmitt Bf 110 day\night fighters over England in 1941; below: Crew members of a Heinkel bomber with ground crewmen on their French base prior to a mission over England.

be ready to take off at first light on the 3rd.

The Lion Geschwader crew was made up of Fw Hermann Wilms, pilot; Uffz Karl Missy, wireless operator/dorsal gunner; Uffz Peter Leushake, observer; and Uffz Johann Meyer, mechanic/ventral gunner. When they were awakened at two a.m. on the 3rd, a big German snowplow and more than 100 soldiers were already clearing the three feet of accumulated snow on the Schleswig runway. The Heinkel crew soon turned out to help in the clearance. The latest report on the British convoy showed it to be heading south off the northeast English coast. The mission that day was to proceed to the convoy, intercept and attack it, and then shadow it to report its position and condition. It was a planned five-hour flight.

At Acklington, Townsend and the pilots of B Flight catnapped in their dispersal hut on the airfield. At dawn the Hurricanes were at readiness. A first radar plot was registering on a cathode-ray tube at Danby Beacon Radar Station. It was 9:03 a.m. The indication was some sixty+ unidentified aircraft approaching the English coast at a height of 1,000 feet. The plots were being read to Fighter Command headquarters and were relayed from there to 13 Group and then Acklington Sector station. Then the phone rang in the 43 Squadron dispersal hut. "Sector Ops here. Blue Section. Forty-three. Scramble base. Angels one." In three minutes Townsend, F/O Tiger Folkes, and Sgt Jim Hallowes were climbing from the Acklington runway when their radios squawked "Vector one-eight-zero. Bandit attacking ship off Whitby. Buster." Racing at full throttle, the three Hurricanes roared over the wave-tops in spread-search formation with Hallowes to the left of Townsend and Folkes to his right. Townsend was the first to spot the Heinkel, above and to the right under cloud. He banked hard right in a climbing turn. "Then I was firing. It never occurred to me at the time that I was killing men. I saw only a big Heinkel with black crosses on it. But in that Heinkel Peter Leushake was already dead, and Johann Meyer, his stomach punctured by bullets, mortally wounded. Closing in fast on the Heinkel, I passed it as it entered cloud—a vague black shadow uncomfortably close above. Then Folkes, the Heinkel and I tumbled out of the cloud almost on top of one another. And the German turned shoreward with a trail of smoke behind him." Karl Missy had been shot in the legs and back. He continued firing his gun as Wilms fought to keep the damaged bomber in the air. He brought it over the high cliffs at Whitby, roaring low over the houses.

Karl Missy watched as the snow-covered ground, houses and trees grew larger, as did the many people watching the Heinkel ripping through telegraph wires, and then they hit and the big plane slithered across the snow heading straight for a barn. Townsend: "I too watched them from where I was circling a few hundred feet above. I could see snow and mud flying up behind the Heinkel as it careered across the ground toward a line of trees. Its right wing hit one, snapping it in half. Then it slewed around and came to rest a few yards from Bannial Flat Farm."

As local police constable Arthur Barratt arrived at the crash site, he saw Wilms kneeling in the cockpit area, burning the aircraft's papers. Some farm workers came up to the wreck and Barratt told them to look after Wilms while he and the others brought out the other crewmen. Wilms then managed to set fire to the plane. The small crowd that had gathered soon put out the blaze with fire extinguishers and shovels of snow.

When Karl Missy tried to get off his swivel seat under the gun, he realized how badly he had been wounded. His legs were shattered and he had to use his arms alone to lower himself to the floor of the plane, near where Johann Meyer lay in agony with several stomach wounds. Missy tried to help his friend, but instead collapsed on top of him. He was losing a lot of blood, but remained conscious and yelled to Wilms to come and help him with Johann. He managed to drag himself out onto the wing where he was able to slide down to the ground. Wilms dragged the body of Leushake out of the plane.

A Mrs Smales and a Miss Sanderson emerged from the onlookers and helped Wilms carry Meyers and Karl Missy into the nearest house. There they wrapped the German crewmen in blankets with hot water bottles and gave them tea and cigarettes. Soon a local doctor arrived and administered morphine to Meyer. Then he cut off Missy's flying boots and put splints on the German's badly wounded legs. An ambulance arrived to take Missy and Meyer to a hospital where that night Karl's right leg was amputated.

The nurse that tended him in the hospital spoke no German and he no English, but he was grateful for the care and kindness she showed him. Then she brought him a visitor. The nursing sister told Peter Townsend that Karl Missy was very ill and she couldn't tell yet whether he would live. Then she held open the door to the ward for Townsend. "I entered and, walking straight up to his bed, held out my hand. Turning to me, he clasped it with both of his until it hurt. But it was the way he looked at me that I can never forget. We had no common tongue so could only communicate as the animals do, by touch, by expression, and by invisible means. As he took my hand Missy had in his eyes the look of a dying animal. If he had died I would have been his killer. He said nothing and only looked at me with a pitiful, frightened, and infinitely sad expression in which I thought I could recognize a glimmer of human gratitude. Indeed Missy felt no bitterness. He sank back on the pillows and I held out the bag of oranges and the tin of fifty Players I had brought for him. They seemed poor compensation.

"Then I left Karl Missy and went back to Acklington and the war. Peter Leushake and Johann Meyer were buried with full military honors at Catterick. A wreath was placed on their coffins. It read, 'From 43 Squadron with sympathy.'"

left: 104-victory German fighter ace Adolf Galland in northern France, 1940; right: Galland in 1967.

Channel Attacks

On 10 May 1940, the squadron commanders at Tangmere on the Sussex coast told their pilots that the Germans had "walked into France today" and we take off at one o'clock sharp for France. And since dawn the Hurricanes of 85 Squadron had been attacking German bombers over Belgium and Holland.

2 July 1940. Until this day Adolf Hitler had fully expected the British government, in the person of the new prime minister, Winston Churchill, to accept the reality of German might, its excellent army, its blitzkrieg armour and artillery, and its superb air force, and in assessing that overwhelming threat, yield to reason and seek an armistice with Germany. Surely, after the recent achievement of the Germans in rolling over Czechoslovakia, Poland, Denmark, Norway, Holland, Belgium and, especially France, the British must see the writing on the wall and accept the inevitable. So confident had Hitler been that the British would cower before the spectre of his armed forces and join the ranks of his conquests that he even partially demobilized the German army.

Two months earlier, a defiant Churchill had told the world "Hitler knows he must break us in this island or lose the war." Now, in the face of a genuine invasion threat by the Germans, Britain was, if anything, more defiant. With the defeat of France in June, the British were truly alone against the German enemy, but incredibly, they chose to resist and fight on against outrageous odds. Indeed, Churchill and his Cabinet had never even discussed the possibility of surrender.

By the 2nd of July, the German leader was persuaded that he no longer had a choice, but had to go ahead with development of a firm and final plan for the invasion of Britain. "Since England, in spite of her hopeless military situation, shows no signs of coming to an understanding, I have decided to prepare a landing operation and if necessary carry it out." His purpose with the operation, called Sea Lion, was "to eliminate the English homeland as a base for the prosecution of the war against Germany." On 19 July, Hitler railed against Churchill in a speech to the Reichstag. He then appealed to the common sense of the British people, telling them it was useless to continue the war, and inviting them to be reasonable and give up the struggle. That evening the London newspapers gave their response, an unequivocal NO!

Hitler was convinced, too, that for Sea Lion to succeed, his air force must gain air superiority over the British Isles. The strutting, arrogant Goering was barely challenged when this requirement was given to him. He boasted about crushing the Royal Air Force in two weeks or less. And, on 1 August Fuehrer Directive No 17 was issued: The Luftwaffe was to overpower the Royal Air Force in the shortest possible time; After achieving air superiority, the air war was to be directed against ports and food stores [London was then the largest port in the world].

The German Chief of the Operations Staff, Generaloberst Alfred Jodl, optimistically wrote to Hitler that "The final German victory over England is now only a question of time." He told the Fuehrer that "all-out air attacks on the United Kingdom—with no holds barred—were not only an indispensable prelude to invasion, but offered the best chance of reducing the affair to an unopposed landing, or even making it unnecessary for troops to go ashore at all. If the Royal Air Force and its supporting industries could be smashed, then Britain would have lost her last weapon, for the Royal Navy would be

The crew of a Dornier bomber after completing a raid on a British target; below: Barges being prepared for Operation Sea Lion, the German invasion of Britain, planned for autumn 1940.

powerless to injure Germany without air support. At the same time, attacks on ports and shipping must not be relaxed. On the contrary, the Luftwaffe must do all it could to hasten surrender by re-doubling its efforts to interrupt supplies."

Wholly confident in the ability of his Luftwaffe to destroy the RAF in a month or less, Goering pegged the start of his all-out air offensive against Britain on the completion of German naval preparations, which would take several weeks. He aimed to begin the major operations at the end of July or in early August. In the interim he intended to use Luftflotten 2 and 3 in an intensive series of bombing attacks on British ports and especially shipping. An added advantage to these attacks would be in aid of the main German air assault through forcing the RAF to wear down their fighter force prematurely in protecting the British convoys.

In the first week of July, Goering's bombers and dive-bombers launched a series of attacks on the British coastal merchant ship convoys in the English Channel and the Straits of Dover, as well as those on their way from the North Sea to the Port of London. There was nothing particularly new for the Luftwaffe in these raids on enemy ports and shipping. They had been undertaking such attacks for more than a year, in the Firth of Forth, at Scapa Flow, and in an organized offensive against the merchant shipping. But the new operations were greatly intensified and, with the French defeat the two main German air bombing fleets were based in France and the Low Countries—much closer to their targets.

During the Battle of France, the RAF suffered the loss of more than 400 Hurricane fighters in a three-week period and were in deep trouble numerically in the Dunkirk evacuation. It was only thanks to an astonishing spurt of productivity by the British aircraft industry operating under the firm hand of Minister of Aircraft Production Lord Beaverbrook, that Fighter Command chief Dowding was able to remain competitive, though he was still short of more than 200 pilots. What, perhaps, mattered most about the Beaverbrook resupply of fighters at that time was that it allowed Dowding the capability to deal with the suddenly increasing pressure being applied by the Luftwaffe on British shipping in the Channel and the North Sea. This was the preliminary phase of the Battle of Britain.

Since April, when Air Vice Marshal Keith Park departed Stanmore where he had been working for Dowding as Senior Air Staff Officer, Fighter Command, Park had taken up a new post in command of No. 11 Group RAF, responsible for the fighter defence of London and southeast England. It was his job to arrange the fighter patrols over Dunkirk during the evacuation—and vitally—to cope with the greatest demands on Fighter Command during the Battle of Britain. Park had a well-earned reputation as a superb tactician and was a popular leader figure as a very much hands-on commander, personally taking direct command of the fighter squadrons on the most important dates in the Battle, Adlertag: 13 August, The Hardest Day: 18 August, and Battle of Britain Day: 15 September. Park had direct responsibility for forty percent of the RAF fighter force and was constantly aware that a significant mistake on his part, or any reckless use of his resources at any point in the Battle could cost Britain the war in a matter of hours.

In the early afternoon of 10 July, a west-bound British convoy sailed southwest through the Straits of Dover, where it came under the escort of six Hurricane fighters ordered up by Park's headquarters to patrol the narrowest part of the Straits. The successful activities of the German U-boats in recent weeks had necessitated the re-routing of nearly all British shipping traffic to ports on the English west coast. However, a number of important local coastal convoys carrying coal and coke had to carry on

left: A Do 17 medium bomber ready for take-off; below: The c of a Heinkel bomber discussing their imminent attack on a Briti airfield in August 1940.

far left: Luftwaffe fighter pilot Horst Petzschler achieved twenty-six aerial victories in the Second World War; left: Oberst Werner Mölders was appointed Luftwaffe Inspector General of Fighters. He died in the crash of a He 111 during a heavy storm; below: The navigator and slightly-wounded pilot of a Heinkel bomber in 1940.

as normal on their original routes and required Park's escort service. As this particular coastal convoy was passing the port of Dover at 1:30 p.m., a nearby radar station began reporting the assembly of aircraft behind Calais and plot activity began on the operations tables of 11 Group and its sector stations, and at its Uxbridge command headquarters.

Park's people immediately ordered a flight of Hurricanes of 56 Squadron up from Manston, near Ramsgate. They had flown over that morning from their base at North Weald in Essex. As the six RAF pilots climbed from the Manston runway they heard from the controller that the convoy was already under bombing attack. In a few minutes they sighted the German bombing force and its large fighter escort stacked up high in three distinct formations. The low group was twenty Dornier Do 17 light bombers. Just above them were a group of Messerschmitt Me 110 close escorts and above them a top guard force of about twenty Messerschmitt Me 109 fighters.

As the Hurricanes approached, the Me 110s formed into the defensive circle that was their typical tactic in such combats, leaving the 109s to do the bulk of the fighting. Hugely outnumbered, the Hurricanes roared in to engage the enemy aircraft. Three of the RAF pilots went in after the bombers. In the action, all six Hurricanes engaged with the 109s at one point and, moments later when the enemy aircraft broke off and headed towards their French bases, three of their number had been shot down by the excited Hurricane pilots. Soon the RAF planes were back on the ground at Manston, one of them having to crash-land there, but the pilot was not seriously hurt. The elation of the Hurricane pilots after the day's action was tempered considerably, however, by the realization that evening at Stanmore when reports from all the fighter groups had been received, that Fighter Command had flown more than 600 sorties that day, roughly twice the average daily effort since Dunkirk. For the results achieved it was an expensive day in terms of flying time and wear and tear on the equipment and personnel of a relatively small force desperately needing to save itself as much as possible for the big battle that everyone knew was coming. But Park and Dowding had to do what they could to cope with the demands of the Admiralty for fighter escort to protect the essential convoys steaming with increasing frequency along the English coasts.

On 11 July, at six a.m., Peter Townsend, flying with 85 Squadron, took off in a Hurricane from Martlesham Heath on a lone patrol of the nearby sea lanes. Ascending from the early ground mist into drift-

left: The Focke-Wulf Fw 190 single-seat fighter-bomber first appeared in large mumbers over France in 1941. Faster than any Allied fighter then in service, the 190 was also more heavily armed; right: The clean design and efficiency of the cockpit and panel layout of the Fw is apparent in this excellent view.

ing raincloud, he was soon told by the controller to head seaward on a vector towards "a bandit in the vicinity."

Passing through 8,000 feet and still climbing, he spotted an aircraft heading the other way through the thickening cloud cover, and identified it instantly as a Dornier Do 17 light bomber. Reacting quickly, he yanked the Hurricane around and began to shadow the enemy plane from behind and below, hoping he had not been seen by the German crew. As he edged closer to the Dornier, the heavy cloud

Shot down over the English countryside during the Battle of Britain, this Bf 109 fighter is inspected by British Army personnel.

and rain made it difficult to keep his target in sight. He judged that a few more seconds would put him in firing position and pulled the hood back to get a better look at the foe. Thirty years later, Townsend would meet Uffr Werner Borner, the wireless operator of the German bomber, and learn from Borner that the Dornier had been on the way back to its base that morning when Borner suddenly saw the Hurricane approaching from behind and shouted over the intercom "Achtung, Jäger!" as he began shooting.

Townsend: "I waited a few more seconds while closing to point-blank range, then opened fire. Borner would never forget (any more than I) the crisscrossing of our bright red tracers. He could actually see me in the cockpit."

Borner recounted to Townsend years later how two of the Dornier crew members had been hit by the Hurricane bullets, and there was blood everywhere in the interior of the bomber. A near miss to the pilot's head had starcracked his windscreen and then another bullet struck Borner's gun, knocking it from his hands. Before that, however, the German scored hits on Townsend's fighter. ". . . a bright orange explosion in the cockpit momentarily blinded me. My engine was hit, and the last Borner saw of my Hurricane was as it disappeared in the clouds, streaming black smoke. By some miracle I was not hit myself. I called the controller: 'I'm bailing out. One, two, three, four, five. Please fix my position."

Townsend left the stricken fighter and made the long parachute descent to splash down in the sea about thirty miles from the English coast. There he floated around awaiting rescue which came thirty minutes later when the trawler Finisterre hauled him aboard. Warming him with a tot of rum, the crew of the trawler delivered Townsend safe and sound to dry land at Harwick and that same evening he was flying patrol again from Martlesham.

In their postwar conversation, Borner told Townsend that the Dornier had barely made it back to their base at Abbeville northwest of Amiens in northern France. After crash landing, the crew counted 220 bullet holes in their bomber.

Two weeks of patrols protecting the convoys had seen eighty-five German aircraft downed for the loss of forty-five RAF fighters; a seemingly impressive tally for the British, but Dowding worried whether his comparatively small resources in planes and especially pilots could somehow be maintained and marshalled for the full-scale offensive that was soon to begin. The win-loss ratio was impressive in his favour, but there had been some profoundly difficult days in that period. On 19 July, six of the nine Boulton Paul Defiant interceptors of 141 Squadron were lost to enemy aircraft. The London papers, of course, continued to celebrate the sometimes exaggerated victories of the RAF over the Germans in the Channel actions, but Dowding and Park knew how tenuous their own position was as they were forced daily to spend the strength of their squadrons ahead of the looming real challenge.

The pressure on them continued on the 24th when Kesselring's air fleet made simultaneous attacks on shipping convoys in both the Dover Straits and Thames Estuary. Keith Park's forces were once again heavily outnumbered by those of the enemy, but again they prevailed thanks in large part to his good management of the squadrons at his disposal. Ever conscious of the many critics of his policy of always retaining some aircraft and pilots in reserve rather than throwing everything into the fight, he stuck with his basic approach and the results proved him right. It was mostly a guessing game, for no one in the Royal Air Force at any level could ever know for certain which of the many raids being tracked on the operations tables was the most important and therefore the most worrisome. Frighteningly, no one could tell when or where the big one would occur. Somehow, though, at the end of most days in this early stage, the results continued to favour the British.

25 July. In the late morning a twenty-one-ship convoy of colliers proceeded from Southend. They were escorted by two trawlers and six Hurricanes and were being stalked by a German reconnaissance aircraft as they were passing Deal. The R.D.F. radar stations in the area were picking up indications of an aircraft formation assembling across the Channel and the reports to 11 Group control caused five Spitfires of 54 Squadron to be launched from Hornchurch to patrol the Dover-Deal area.

An initial wave of approximately thirty Junkers Ju 87 Stuka dive-bombers in the company of a large formation of Me 109 fighters, now arrived over the convoy and began their attack. Once again Park

The pilot and gunner of this Messerschmitt Me 110 day/night ground-attack and reconnaissance fighter are assisted by German personnel from the wreck of their aircraft brought down over France by enemy fighters.

was extremely limited in the number of defending fighters he could send up against the German raiders. The slow, ungainly Stukas were easy meat for his Spitfires and Hurricanes, but the small number of fighters he was able to spare for this action, as usual, left them greatly outnumbered against the escorting 109s. Park scrambled some Hurricanes of No 111 Squadron from their forward airfield at Hawkinge near Folkstone, but the time they needed to climb, reach and engage the German aircraft negated whatever advantage they might have in the encounter. Dowding, too, was then under considerable criticism for not putting his forward airfields such as Hawkinge to greater use in that period.

In the attack, the Stukas managed to sink five of the colliers and heavily damage a further six, leaving them crippled in the Strait. Many of the Stukas were shot down in the raid, but the results of the bombing so emboldened the Germans that they determined to destroy the remainder of the convoy by sending out E-boats in an audacious daylight attack. In response, the Royal Navy arrived on the scene with two destroyers which scattered the E-boats, but the destroyers were severely shelled by German shore batteries and were finally recalled. As they sailed back to base they were attacked by Stukas and one was heavily damaged. That evening a number of E-boats arrived in the area in darkness and sank three of the crippled colliers.

The devastation to that convoy forced the Admiralty to temporarily suspend such merchant sailings through the Straits pending development of new arrangements for getting such convoys through the dangerous passage under cover of darkness and for providing more effective escort by day. With the convoy activity stood down for the time being, Kesselring was ordered by Goering, despite the present lack of convoy activity in the area, to continue attacking by inflicting maximum damage on warships of the Royal Navy preparatory to the German invasion of England. Kesselring promptly sank two RN destroyers and severely damaged a third, which action caused the Admiralty to admit it had lost command of the Straits by daylight. It recalled its destroyers from Dover and withdrew them to Portsmouth.

In the late afternoon of 7 August, a new, entirely revitalized west-bound Channel convoy was assembled in the Thames Estuary. It was being escorted by two newly rearmed destroyers with improved anti-aircraft capability. The convoy was also being escorted by specially-modified vessels flying barrage-balloons for further protection against the enemy dive-bombers. The plan for the convoy called for it to be met at first light the next day a large RAF fighter escort. In the dark of that night, however, an unseen flotilla of enemy E-boats crept in on the merchant vessels of the convoy as it entered the Straits of Dover.

Early in the pre-dawn hours of the 8th, the E-boats attacked the ships of the convoy along the Sussex coast creating extreme chaos. Two of the merchant ships were sunk, with a third damaged, and in the confusion two merchant ships collided, raising the German tally to three sunk and two damaged. What now remained of the convoy was no longer a formation, but individual vessels scattered over a vast area of sea. There was no convoy for Park's fighters to escort. At least six of the ships, including one sunk in the collision, were lost.

This early phase of the Battle of Britain was drawing to a close. Hitler had told Goering on 30 July to be ready to launch the main assault on twelve hours' notice. The Luftwaffe chief was not quite ready to go yet, however, and took advantage of questionable weather predicted for the coming week before alerting his commanders to prepare for the start of the full-scale offensive on 10 August; it would in fact be postponed until the 13th, Eagle Day.

Radar Chain

The Chain Home radar system was made up of twenty-one coastal stations with 300-foot-high steel lattice masts. These vulnerable structures could "see" out across the sea over an arc of 120 degrees and detect aircraft at more than a hundred miles distance and, excepting very low-flying examples, read their height and direction of travel. The detection capability was seaward only; once an aircraft crossed the cost and headed inland it could no longer be tracked by the system. They could determine the range, height, bearing, and strength of a hostile raid. Once inland though, the enemy raiders had to be plotted by the members of the Observer Corps, by sight and/or sound, depending upon visibility. The Observer Corps posts covered the country and depended on the eyes, ears, and field telephones of the volunteer members to provide a vital link in the air defence/detection system of Fighter Command by day, at night, and in all weathers.

When the Chain stations picked up radar plots they were immediately passed to Fighter Command at Bentley Priory, Stanmore, where they came into the Filter Room and a filter officer checked them against information about friendlies and hostiles before passing them on to the adjoining Operations Room, which was Dowding's control centre. There, on a twenty-four-hour basis, he could view on a large map table the exact location in the British Isles of a developing raid or one (or more) in progress. The room was staffed by the commander's duty controller, and liaison officers in contact with the Observer Corps, Bomber and Coastal Commands, Anti-Aircraft Command (guns and searchlights), the Admiralty, and the Home Office (air raid warnings).

The principal Fighter Command groups were responsible for limited areas. For example, 11 Group covered an area from Portsmouth to the Thames Estuary; 12 Group north from the Thames Estuary to Yorkshire (the Midlands and East Anglia); 10 Group covered southwest England; and 13 Group the north of England and Scotland. The filtered radar plots were sent by Fighter Command to the fighter groups which also received Observer Corps plots direct from the Observer centres. Those running the balloon barrages were also in direct contact with the Group Operations rooms where the Group Commanders and their controllers could all simultaneously see the actual up-to-the-minute situation on the table maps in their Ops rooms. The table maps were "manned" by WAAF girls wearing headphones and wielding croupier rakes with which they moved coloured discs representing aircraft per the plots they received. The group controller in each group directed the appropriate sector station to handle each raid. The sectors represented geographical areas within each group. Of the various airfields located within each sector, one was designated the Sector station and operated its own Operations Block which housed the Ops Room proper and a D/F (Direction Finding) Room. The personnel manning the D/F Room tracked the sector's own fighters, using radio "fixes" obtained from IFF (Identification Friend or Foe), and also referred to as 'Pip Squeak', an automatic transmitter fitted in all Fighter Command aircraft.

The operators in the D/F Room passed the fighter positions to the two deputy controllers in the Ops Room where two navigators immediately determined the fighters' interception courses, passing them to the controller whose elevated dais provided him with an excellent view of the table map of his sector and the coloured discs showing the positions of the friendly and hostile aircraft, as positioned by the WAAFs with the rakes. The sector controller then directed his fighters by accurate vector to intercept the hostile raiders, using a brief code: Scramble (take off); Angels Ten (height 10,000 feet);

far left: Air Chief Marshal Hugh Dowding, who headed RAF Fighter Command in the Battle of Britain period; left: Assisstant Section Officer Edith Heap was a WAAF plotter and later an Intelligence Officer; below: The plotting table in the underground Ops Room of No. 11 Group, Fighter Command, at Uxbridge, November 11, 1942.

RAF Hurricane interceptors climb from their English base to engage the enemy raiders approaching from the French coast; below; an Observer Corps member at work, summer 1940.

Orbit (circle a given point); Vector One-Eight-Zero (steer course 180 degrees); Buster (full-throttle, expedite); Tally-ho (enemy sighted); Pancake (land); Bandit (enemy aircraft).

Code-names were devised for each sector station and each fighter squadron in two-syllable words, examples being 'Jaunty' and 'Lumba.' With squadrons typically operating as twelve-aircraft tactical units, each was divided into four sections of three aircraft, with Red and Yellow in A Flight; Blue and Green in B Flight; the pilots in each section were identified as 1, 2, and 3 (i.e. Red 1 led Red Section, followed by Red 2 and Red 3).

The organization of this elaborate communications system worked as follows: from hostile target to radar station by radiomagnetic waves; from radar station to Fighter Command by telephone line; from Fighter Command to Group and Sector and from Sector Ops to the fighter pilot in the air by radio / telephone.

For years both the British and the Germans had been working independently on their own version of military radar. The British were primarily, perhaps even exclusively, interested in developing the technology for the defence of their island nation, to link coastal radar stations by telephone and radio / telephone to their defending fighter aircraft. Their long-range radar, while less sophisticated than German radar, was effectively integrated into a sound, practical air defence system. Freya, the German long-range radar, was capable of detecting hostile aircraft to a range of seventy-five miles and through a 360 degree circle, though it could not detect altitude. And, unlike the British radar which incorporated gigantic fixed steel lattice masts, Freya was mobile. In their Wurzburg short-range radar, the Germans had developed a relatively small, completely mobile apparatus for accurate detection, especially in aid of anti-aircraft guns and searchlights. It was the Wurzburg system to which he was referring when Hermann Goering famously remarked, "If a single enemy bomber crosses our frontier, you can call me Meier."

The Germans had been curious and intrigued by the continuing construction of huge masts along the south and east English coasts in 1939. Assuming the high masts were part of a new radio system, they decided to learn all they could about what the British were up to in those lengthy construction projects. While they undoubtedly discovered some of what they wanted to know through the use of spies, they also employed a somewhat more elegant approach when they sent the great airship Graf Zeppelin across the Channel, fitted out with cameras and electronic detection equipment for a spot of aerial reconnaissance.

With the Luftwaffe Signals Section chief General Wolfgang Martini on board, the airship cruised leisurely up the east coast in late May, Martini and the crew unaware that they were being plotted throughout their journey by the very radar stations they had come over to observe, and those Zeppelin plots were with the controller at Fighter Command. The reconnaissance mission came to nothing and, in the evening of 2 August, only a month before the onset of war between Britain and Germany, the Graf Zeppelin set out again in the darkness, the crew hopeful of making a breakthrough discovery of what the British were up to along those high cliffs. Again, the recce came to nothing more than to kindle the amusement of two patrolling RAF fighter pilots and that of the coast guardsmen near Aberdeen.

Quite soon, the British Chain Home radar system would be completed and would figure dramatically in the outcome of history's first great battle in the air.

Pilots of 19 Squadron on their Duxford, England base in 1940.

Targetting The Airfields

At a few minutes after seven in the morning of Tuesday, 13 August 1940, Brian Binyon, a controller with the Bromley Observer Corps, spoke calmly into his field telephone as he reported what he was seeing: "Raid 45 is bombing Eastchurch drome." Binyon watched as the fifty+ bomb aimers of Luftwaffe Bomber Group 2 delivered their bombloads on the RAF airfield. Across the field, in the Non-commissioned Officers' quarters, Sgt. Reg Gretton of 266 Squadron, recovered from momentary shock and yelled: "They're dropping bombs! They're dropping bombs on us!" The more pragmatic commanding officer, Group Captain Frank Hopps reacted more analytically: "My God, this station's worth millions. Some accountant's got a job to do writing off this lot."

Oberst Johannes Fink, the German air leader in command of this early morning attack on the airfields of Fighter Command, was pleased and impressed with what his group was achieving this morning. To that point, twelve parked Spitfires that had been on overnight transit to RAF Hornchurch in Essex, had been destroyed, along with five Bristol Blenheim light bombers. The Germans had also left the Operations Block severely damaged with all electrical and telephone lines cut, and nearby petrol supplies destroyed.

The surprise felt by Sgt Gretton and Group Captain Hopps was rooted in the fact that Eastchurch was in no way connected with Fighter Command or its airfields. It was a Coastal Command field, whose squadrons were responsible for patrolling the North Sea in search of German naval raiders, an unlikely target for enemy bombers set on laying the groundwork for the invasion of England.

Having shifted their attention from bombing the British coastal shipping, the Germans were now concentrating their considerable efforts on the task of destroying the Royal Air Force, initially its airfields and specifically, those of Fighter Command. In the late afternoon of the 13th, eighty-six Stuka dive-bombers broke up the tea break of the airmen at RAF Detling near Maidstone, Kent. In the attack the main runway was cratered, the hangars were set alight, the Ops Block demolished and upwards of twenty aircraft were destroyed on the ground. Devastatingly, some fifty personnel were killed. The base was out of operation for a day, in yet another example of Luftwaffe intelligence and planning seemingly gone astray, for Detling, like Eastchurch, was a Coastal Command air station and nothing to do with Fighter Command.

But these attacks, and the largely ineffectual earlier raids on the radar stations of the RAF, were soon overshadowed by such recent raids as that of Luftwaffe Test Group 210 on the important Fighter Command forward airfield at Manston. The Germans struck the base, code-named Charlie Three, with a large combined force of bombed-up Me 110s and Me 109s under the command of Hauptmann Walter Rubensdörffer. His fighters arrived in time to catch the Spitfires of 65 Squadron struggling to get into the air from the Manston drome as they taxiied blindly through the thick, choking smoke of the many explosions. Flight Lieutenant Al Deere, in the lead of 54 Squadron, recalled being over the Manston base during the attack: "... there was a cloud like white pumice rising over the drome ... it was like a shroud over everything." What Deere was seeing was chalk dust blown up from more than a hundred bomb craters. The same sight greeted Flying Officer Wilfred Duncan Smith, father of the current British government Work and Pensions Secretary, Iain Duncan Smith, when he arrived in the Manston area after returning in a Tiger Moth biplane from leave.

The Manston base was an all-grass airfield built in 1916 during the First World War. In many ways,

above: Airmen of the Royal Air Force being given physical training in their preparation to become new pilots; at right: The pilot of Luftwaffe Heinkel He 111 bomber bringing his bombload to Britain during the Blitz period in early 1941.

in 1940 the station was still operating as it had in the First War period, by the book and keeping largely peace-time hours. On one occasion, some pilots of 32 Squadron were refused transport to the mess when they couldn't produce a signed Form 658, so they appropriated a tractor at gunpoint and, on arriving at the building, discovered that the chef had already left for home, having locked up the food larder. The squadron's John Worrall: "I shot the lock off the larder and we ate."

A lot of Manston personnel were, perhaps, more nervous about the bombing raid than those of other RAF fighter stations receiving similar Luftwaffe attention that week. At Manston, many of the ground personnel, when the raid struck, headed rapidly for the warren of deep chalk caves that wound beneath the aerodrome. Hundreds of base workers holed up in the catacombs for days, despite the efforts of their officers to dislodge them.

As the German fighter leader Adolf Galland recalled: "The enemy air force must be wiped out while

still grounded" as he considered the clinical elimination, airfield by airfield, of Fighter Command's facilities. What was becoming clear, though, was that Abteilung 5, the Luftwaffe's Intelligence arm, was not distinguishing between the airfields of Fighter Command and other military airfields in southern England, in its contributions to target selection. Many wondered about the logic of the target selection. Which of the Fighter Command airfields were of highest priority for attack, and when attacking them, which specific facilities mattered most as targets—the hangars and other buildings, or the aircraft on the ground?

Another forward fighter base came under Luftwaffe attack on 12 August. RAF Hawkinge near Folkstone, right on Kent's coastal cliffs, was pounded by Junkers Ju 88 bombers and put out of action for the day. The field was hit again three days later by Stuka dive-bombers and effectively shut down for two further days. In that period, the Germans struck again at Eastchurch, and at the Short Brothers

Aircraft factory at Rochester in Kent, another questionable target in that the Shorts plant there was producing four-engined Stirling bombers, rather than front-line defensive fighters, the supposed prime target of the Luftwaffe in its preparatory work towards the invasion plan. Other questionable raid targets in that brief span were the naval air stations at Lee-on-Solent, Ford, and Gosport, none of them with any Fighter Command connection.

Between 12 and 23 August, the second phase of the Battle of Britain was conducted when bombers and dive-bombers of the Luftwaffe, accompanied by large numbers of fighter escorts, targetted the key coastal and near-coast airfields of RAF Fighter Command in a major attempt to destroy the main

far left: A German aerial view of the RAF airfield at Biggin Hill, Kent, during the Battle of Britain; centre left: Pilot's view from the cockpit of a Spitfire; left: Pilot Officer R.F.T. Doe who flew with Nos 234 and 238 Squadrons in the Battle of Britain and after; above: A Vickers-Supermarine Spitfire Mk I fighter which, together with its sister aircraft, the Hawker Hurricane, were the principal RAF fighter aircraft of the Battle of Britain period.

defensive capability of the British air force and gain air superiority over Britain ahead of the planned German invasion.

Gradually that week, the Luftwaffe seemed to get the measure of the target airfields they were trying to destroy. 15 August, though, saw the downing of seventy-five German aircraft in the course of their raids, a momentous achievement for Fighter Command in the midst of the punishing attacks on its airfields. On the 16th, however, the Germans were back in force to continue their visits to the fields of Fighter Command as two squadrons of Ju 88 bombers hit one of 11 Group's most important sector stations, West Malling, Kent at about 11:00 a.m. Within the hour, the Chain Home RDF system detected three large formations of aircraft heading towards the Thames Estuary. They would be met by upwards of eighty of Keith Park's fighters which turned many of the bombers back to France. Almost simultaneously, a force of about 150 German planes crossed the south coast unopposed between Folkstone and Brighton. By the time fighter defenders reached the intruders, the Germans were bombing Farnborough and the London docks where, in the attack, sixty-six civilians were killed. By 12:45 p.m. Park had scrambled eight fighter squadrons to intercept a new formation of 150 enemy aircraft coming across from Cherbourg, consisting of many Stukas, Ju 88s, and escorting 109s. As the massive formation approached the Isle of Wight, it split into three elements. One headed for

the Ventnor Chain Home station, a second for Gosport, and the last for RAF Tangmere. Given sufficient warning the Hurricanes of 43 and 601 Squadrons at Tangmere were scrambled to intercept the enemy force over the Solent. In about eight minutes of aerial combat, some of the estimated fifty to 100 Ju 87 Stuka dive-bombers turned back to their French bases, jettisoning their bombs as they fled. In the ensuing action, the large escorting force of Me 109 fighters took no part in the engagment. In the fight, which soon broke into a number of individual fights, 43 Squadron pilots Tony Woods Scawen was slightly wounded and was forced to crash-land at Parkhurst on the Isle of Wight, while Hamilton Upton had to force land on the beach at Selsey.

As this air combat continued, the first wave of Stukas went on to strike in the first attack of the war on the airfield at Tangmere, near Chichester on the Sussex coast. Meanwhile, 601 Squadron, also out of Tangmere, was vectored over the Isle of Wight and told to climb to 20,000 feet. On the way they spotted the wave of Stukas headed inbound towards Tangmere. The 601 leader, Flight Lieutenant Archibald Hope, then disregarded the controller's instructions and took the squadron down after the Stukas, which were now beginning the dive-bombing of the airfield. Pilots of 601 soon shot down three of the Stukas, but not before they had dropped their bombloads on the base. In the encounter with the Stukas, the Hurricane of American Pilot Officer Billy Fiske, flying with 601, was hit by gunfire

below: Stan Turner flew with No 242 (Canadian) Squadron, RAF in the Battle of Britain; far left: Twenty-two-victory ace Wing Commander Robert Stanford Tuck flew Hurricanes with 257 Squadron and was later a prisoner of war at Stalag Luft Three.

from one of the German planes. The burning Hurricane was trailing glycol as Fiske managed to put it down on the airfield. Two nursing orderlies, Corporal George Jones and AC2 Cyril Faulkner quickly arrived in an ambulance to pull the injured Fiske from the wreck of his fighter and take him to the station sick quarters. Dr C.B.I. Willey: "I saw one of 601's Hurricanes lying on its belly belching smoke after coming if for its final approach. There were two ambulance men there. They had got Billy Fiske out of the cockpit. They didn't know how to take off his parachute so I showed them. Billy was burnt about the hands and ankles. I told him, 'Don't worry, you'll be all right...'"

At that point, Dr Willey heard a warning over the station Tannoy: "Take cover! Take cover! Stukas sighted coming towards Tangmere. Take cover!" He got his twelve patients moved quickly to a bomb-proof shelter but was then injured himself when a bomb struck the sick quarters. Dr Willey carried on treating the wounded and seriously injured and, when Billy Fiske was brought in, he examined the pilot, who had been badly burnt from the waist down. Fiske was unconscious and the doctor gave him morphine. Within twenty minutes Fiske was taken to the Royal West Sussex Hospital in Chichester where he died the next day.

Westhampnett was a satellite airfield of Tangmere, also near Chichester. There the Spitfire pilots of

all: Photos representative of the Spitfire manufacturing process at the Vickers Castle Bromwich factory in Birmingham, England in 1942. At Castle Bromwich, and at many other aircraft manufacturing facilities across Britain in the Second World War, women played a vital part in advancing the quantity and quality production of the superb fighters and other aircraft for the pilots and aircrews of the Royal Air Force and Royal Navy in the war years. right: Cockpit layout installation.

above: Inspection of a nearly completed Spitfire on the Castle Bromwich assembly line in 1942; left: Finished Spitfires awaiting pre-delivery test flights. At Castle Bromwich, chief test pilot Alex Henshaw probably flew more Spitfires than anyone in the history of the remarkable plane; above right: RAF Fighter Command pilots put some muscle into pushing a new Spitfire at their fighter station dispersal in 1942.

602 Squadron were ordered into the air just before 1:00 p.m. to attack the Stukas at Tangmere. Findlay Boyd, a flight commander with the squadron, spotted and went after a Stuka that was climbing out after dropping its bombs on the airfield. He chased it and soon shot it down. Three other 602 pilots caught and downed Stukas departing from the attack. On the ground at Tangmere, 2nd Lt. E. P. Griffin of the Royal Engineers Construction Company, manned a Lewis gun on the airfield and managed to shoot down a Messerschmitt Bf 110 twin-engined heavy fighter which crashed nearly a mile from the drome, killing its three-man crew.

The twenty-minute German bombing attack on Tangmere had begun at 1:00 p.m. Destroyed in the raid were all of the pre-war hangars, the station workshops, the water-pumping station and stores. Badly damaged were the Officers' Mess, and the power, water, sanitation and communications systems. Seven Hurricanes, six Blenheims, and a Magister aircraft were destroyed on the ground, along with forty vehicles. In the raid, twenty-five German aircraft were brought down. Ten Tangmere station personnel were killed and twenty injured.

At lunchtime on 18 August, thirty-one Dorniers of Oberstleutnant Fröhlich's Bomber Group 76 had been sent to strafe the Fighter Command sector stations at Biggin Hill in Kent, and Kenley in Surrey. As a diversionary tactic, the Luftwaffe had also sent five Dorniers in at low level to confuse the radar stations. It immediately became a split-second decision for the Kenley controller, Anthony Norman; though the Observer Corps had sighted the low-level Dorniers as they approached the white chalk quarry that pointed to Kenley. 11 Group had not ordered a scramble, so Norman acted on his own. "Get them into their battle bowlers . . . tin hats everybody," he ordered the floor supervisor. To Squadron Leader Aeneas MacDonnell's 64 Squadron he ordered: "Freema Squadron, scramble. Patrol base, angels twenty."

High over Kenley, MacDonnell suddenly told his pilots, "Freema Squadron, going down." One of them, Sgt. Peter Hawke recalled: "Why down? We need all the height we can get." Then he saw the black smoke column pulsing up from the Kenley hangars. The low-level raid had come in ahead of the attackers. Seeing a flash like exploding helium from a Dornier, Hawke thought, "My God! Did I do that? Well, this was what I was trained to do."

Just over the airfield, Hurricanes of 111 Squadron were arriving from Croydon, but no one had warned them that the station defences were firing parachute-and-cable rockets at the German raiders. The electrically-fired rockets were snaking upwards at forty feet a second to grapple the enemy wings with steel wire. "If one of those hits us we're finished," thought the 111 commander, Squadron Leader John Thompson.

In the attack, Bomber Group 76 lost six Dornier bombers and crews and four Ju 88 bombers. It had smashed ten hangars and damaged six more; put the Ops Room out of action, and crumbled many more buildings. Fortunately for Kenley, many of the enemy bombs failed to explode.

Group Captain Richard Grice at Biggin Hill down the road from Kenley knew how lucky Biggin had been when its turn for the attentions of the Luftwaffe had come. Most of the 500 or so bombs that fell on his airfield had landed wide, out on the eastern end of the field. Still, he warned the station personnel: "What happened to Kenley today can well happen here, so don't think that you've escaped."

At Fighter Command, Dowding's former personal assistant, Pilot Officer Robert Wright remembered wondering: "Did the Germans plan to concentrate the might of their bombers against the sector stations? German General Paul Deichmann, Chief of Staff of the 2nd Flying Corps, later recorded: "That fear was groundless. Never at any time did the Luftwaffe High Command suspect that Kenley and Biggin Hill—or, for that matter, Hornchurch, Tangmere, and Middle Wallop—were sector stations, the nerve centres of Dowding's command. We all thought priority comnmand posts would be sited underground, away from the centre of operations, not in unprotected buildings in the centre of the airfields. And not all of them had sandbags or blast walls!"

By 24 August all efforts to hold Manston had proved in vain. As twenty Ju 87 dive-bombers with fighter escort swept in over the field, Pilot Officer Henry Jacobs was relaying a blow-by-blow commentary to HQ 11 Group, when he heard a hollow note like a gong going up the wire; a bomb, striking the telephone and teleprinter links, had severed 248 circuits in one blow. Running from 600 Squadron's Ops Room, Jacobs saw that the East Camp guard house had disappeared, swallowed into a chalky crater forty feet deep. All the Manston buildings still standing were burning. That same day most of the station personnel were moved out. Already civilians were moving in to loot tools and live ammunition from the main store.

Interviewed at the International Military Tribunal at Nuremberg in 1946, former Luftwaffe chief Hermann Goering: "I believe this plan [raiding RAF airfields and British aircraft factories] would have been very successful, but as a result of the Fuehrer's speech about retribution, in which he asked that London be attacked immediately, I had to follow the other course. I wanted to interpret the Fuehrer's speech about attacking London in this way. I wanted to attack the airfields first, thus creating a prerequisite for attacking London … I spoke to the Fuehrer about my plans in order to try to have him agree I should attack the first ring of RAF airfields around London, but he insisted he wanted to have London itself attacked for political reasons, and also for retribution. I considered the attacks on London useless and I told the Fuehrer again and again that inasmuch as I knew the English people as well as I did my own people, I could never force them to their knees by attacking London. We might be able to subdue the Dutch people by such measures but not the British."

The famous pilot's signature board from the White Hart pub, Brasted, near RAF Biggin Hill in Kent.

Bombs For Britain

It was a sunny Saturday, 7 September. The people of London were mostly tending to their normal weekend tasks, accelerated somewhat by the pressures of wartime. The more fortunate were relaxing in the warm sunshine of Hyde Park and St James Park. Some were enthralled with an afternoon cricket match, showing their occasional approval with gentle, mannerly applause. Over Canterbury in Kent, Heinkel He 111s of Bomber Group 2 under the command of Oberst Johannes Fink roared over the cattle market. It was 4:30 p.m.

Few, if any, in London that afternoon had reason for particular concern about what the enemy might be up to that fine summer day. Most were comforted and uplifted by the inspiring efforts of the pilots of RAF Fighter Command who had been doing their best all summer to sweep the skies of southern England clean of those nasty German intruders.

As many Londoners watched, their reverie was suddenly ended. Twenty-one squadrons of Fighter Command Hurricanes and Spitfires had been scrambled and were climbing from their stations at Kenley, Northolt, Biggin Hill and elsewhere, most of them in full expectation that the approaching German bomber formations would be striking again at the fighter fields of the RAF or British aircraft factories like Vickers at Weybridge. What they soon saw was a sight they would never forget, a dark mass of nearly 1,000 enemy aircraft, Heinkel and Dornier bombers and their Messerschmitt Me 109 fighter escorts growling across the green and tan of the English countryside at 8,000 feet and occupying 800 square miles of sky. Clearly, their target was London.

What had led up to this unprecedented raid? In the German hierarchy, General Bruno Lörzer, commander of the 2nd Flying Corps, and Generalfeldmarschall Albert Kesselring, commander of Air Fleet 2, had long been pressing for all-out air attacks on the British capitol. Adolf Hitler refused to allow such an assault, however, in the persistent hope that a "peace" could still be reached with Churchill. But in the evening of 24 August, a small number of Luftwaffe bomber crews committed a slight navigational error and the bombs they had brought for the oil storage tanks at Thameshaven were instead delivered over central London, the first bombs to fall on London since the notorious Zeppelin raid of 1918.

Churchill reacted with fury, ordering an instant reprisal by Bomber Command, which sent eighty-one Whitley, Hampden, and Wellington bombers to Berlin. With the rather crude navigation methods available to them at the time, only about ten percent of the British bombers found their target that night. They returned to Berlin four times in the next ten days. One thing led to another and on 2 September, Oberst Theo Osterkamp told Major Adolf Galland and the other Luftwaffe fighter leaders at Wissant near Calais that a massed bombing attack might be launched on the 7th. On the 3rd, a no-holds-barred conference was held at The Hague, at which Goering said that the time had come to change tactics and re-direct Luftwaffe resources into a massive assault on London. The only question was—had the resources of RAF Fighter Command been sufficiently depleted to this point, or would the cost to the German bomber fleet be too high? Kesselring championed the view that Fighter Command was finished, as evidenced by the Luftwaffe combat reports. He took satisfaction in reminding the gathering that he had all along advocated a mass attack on a single key objective, rather than Goering's campaign of divergent targets—ports and shipping, radar stations, airfields, and factories, struck in turn and then capriciously abandoned. Hugo Sperrle disagreed with Kesselring, sighting the continuing accomplishments of Fighter Command's pilots and claiming that the RAF still retained

left: German raiders taking off to strike at a British target in summer 1940; below: A Heinkel medium bomber is photographed over the river Thames in London during the Battle of Britain.

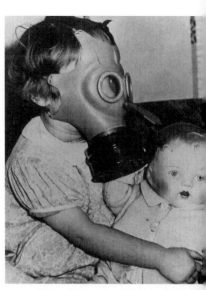

The terrors and deprivations of the German Blitz on London and the other cities of Britain are expressed in the civilian population who lived in constant fear of poison gas attack as well as the devastation wrought by the bombing raids that persisted for fifty-seven consecutive nights from 7 September 1940. Many Londoners sought shelter and some respite from the bombing by descending nightly for protection and rest in the tunnels and stations of the city's Underground system.

more than 1,000 operational fighters left. In fact, the total of serviceable Spitfires and Hurricanes then numbered exactly 746. General Kurt Student, the designated commander of airborne troops for the proposed invasion of England, Operation Sea Lion, said later that Goering had told him "The Fuehrer doesn't want to invade Britain." When Student asked him for an explanation, the Reichsmarschall shrugged and said "I don't know. At any rate, there will be nothing this year." On 4 September, Hitler raged to a worshipful crowd at the Berlin Sportpalast "If they attack our cities, we will raze theirs to the ground. We will stop the handiwork of these air pirates, so help us God."

The planning for a mass raid on the 7th began. 1,273 bombers and fighters were earmarked for the day's activities. Goering intended to personally take charge of the assault from a vantage point on the Channel coast. While many would see the attack as a colossal blunder on Goering's part, others, including 11 Group commander Keith Park, would be grateful. "Thank God for that. I knew that the

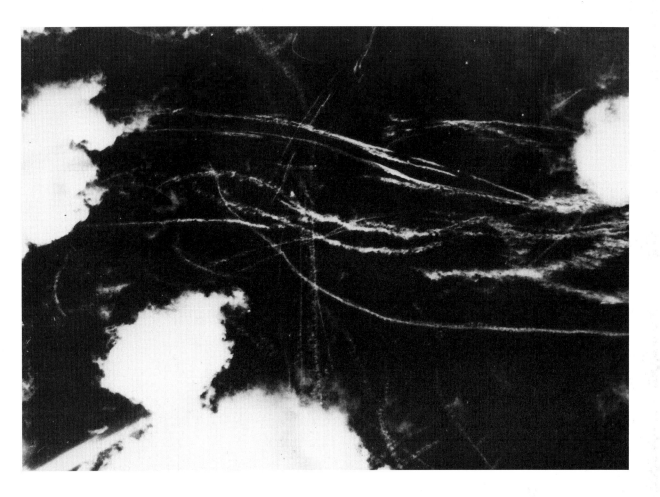

left: One of the defenders of Britain, a Polish pilot flying Spitfires with Fighter Command; above: The common sight of vapour trails high over London and the southeast of England during the Battle of Britain period.

Nazis had switched their attack from the fighter stations, thinking they were knocked out. They weren't, but they were pretty groggy."

In the Thames Estuary, at Gravesend twenty-four miles out of central London, were three veteran American war correspondents who huddled next to a big haystack to watch a hellish scene like nothing anyone had ever witnessed. Ben Robertson, Edward R. Murrow, and Vincent Sheean had, until this day, been covering the Battle of Britain for their news organizations, from the white cliffs of Dover where they had front-row seats for the Luftwaffe attacks on British coastal shipping in the Channel. They had also seen and heard something of the German strikes on Britain's harbours and the airfields of Fighter Command. Londoners were conditioned to the wail of air raid sirens and the thump of anti-aircraft guns issuing from hundreds of batteries in the parks. After a while they paid little heed when the wondrous fighter planes of the RAF rose to tangle with the black-crossed, villainous flying Huns.

above left: A severely damaged Anderson shelter in the gardenof this bombed home; left and above: The preparation of school children for possible evacuation from urban target areas to the relative safety of the English countryside in the days before the beginning of the Blitz.

But now, when the nasty Nazi gnats were starting to show up in wave after wave after wave; in the dark, billowing clouds over the Big Smoke … everything was changing and this time they were bringing the war to the people of Britain, in the start of a fifty-seven night campaign of merciless, terrifying, relentless destruction. The seemingly endless formations of enemy planes, their strangely out-of-sync engines droning on, made their way to the docks of the city's port and soon immense columns of thick smoke were rising there.

It was something less than their finest hour for Keith Park's heroic Few. Massively outnumbered, sent up much too late and much too low, very few of his squadron's were actually given the opportunity to operate at full strength and capability that day. The situation rapidly deteriorated for them, forcing the majority to abandon the tactics they knew so well, and fly as lone wolves. The effective team-work they trusted was out the window. The day would prove an eye-opener for most of them. One, Flying Officer Dennis Parnell of 249 Squadron, found himself playing tag with a burning Heinkel bomber heading back for France. Each time the German plane reappeared to one side or the other

Robert Shaw in a still from the the 1968 film *Battle of Britain*; bottom: actual scenes from the period.

of its heavy smoke trail, Parnell got off a quick burst. He kept that up until the German finally ran out of air and flopped onto the mudflats of Sheerness.

"The sky over London was glorious, ochre and madder, as though a dozen tropic suns were simultaneously setting round the horizon . . . everywhere the shells sparkled like Christmas baubles."
—Evelyn Waugh

The London sky was so thick with twisting, rolling, jinking aircraft of both sides that distinguishing friend from foe was proving a problem for many airmen. When Sgt Cyril Babbage of 602 Squadron caught sight of his friend Andy McDowall being chased by six very determined 109 pilots, he yelled "Hang on, Andy. I'm coming." Babbage had unwittingly alerted a further dozen 109s who also went to work on McDowall, compounding his troubles. Flying Officer Keith Ogilvie of 609 Squadron described his experience: ". . . zooming and dancing around us like masses of ping-pong balls." Ogilvie got in close, took careful aim at one of a pair of the German fighters, and was mortified to discover he had hit the second plane.

One of 222 Squadron's performers that afternoon was Sgt John Burgess, a twenty-year-old Spitfire pilot with a total operational exposure of just ten days at his Hornchurch base in Essex. In the late afternoon that Saturday, Burgess was chasing a trio of Heinkel bombers as they ran hard for their French airfield, having delivered their loads on the London docks. He trailed them by about three miles but, with the considerable speed of his Spitfire he had no concern about overtaking them. At that point Burgess spotted what he believed to be a pair of Hurricanes about two miles behind the Heinkels and apparently also tracking the bombers. He decided to join with the Hurricanes in a concerted effort to down the Germans, but as he drew nearer he saw their yellow noses that marked them out as enemy fighters. Shifting mental gears instantaneously, Burgess pulled into line behind the rearmost fighter and sprayed it with machine-gun fire. The German fighter flipped over and began trailing a thin wisp of white smoke. Burgess and the German fell into a vertical dive at red-line speed. "At that point I suppose I was down to 2,000 feet, but he kept on, straight into the ground. I was very shaken . . . I was shaking because I had obviously killed a man, and I had never killed anyone before."

"I burn for England with a living flame / In the uncandled darkness of the night. / I share with her the fault, who share her name, / And to her light I add my lesser light. She has my arm—who had my father's arm, / Who shall not have my unborn children's arms. / I burn for England, even as she burns / In living flame, that when her peace is come / Flame shall destroy whoever seeks to turn / her sacrifice to profit—and the homes / Of those who fought—to wreckage. / In a war for freedom—who were never free. —from *Poem* by Gervase Stewart

Possibly the greatest danger facing the escorting fighter pilots of Goering's Luftwaffe on the mass raid of the 7th was the Channel itself. Major Max Ibel of Fighter Group 27: "Although the fighters had stuck close to the bombers, it had been a close-run thing. Every warning bulb was glowing red—signalling ten litres of petrol, twenty minutes flying time at most". As they reached the Channel coast of England on the way back to their French bases, the very real prospect of coming down in that water haunted many of them as they stretched their luck and hoped to make it flying on fumes. And now the defending fighters of the RAF were alerted.

"I was pushing the glass across the counter for a refill when we heard it coming. The girl in the corner was still laughing and for the first time I heard her soldier speak. 'Shut up!' he said, and the laugh was cut off like the sound track in a movie. Then everyone was diving for the floor.

"The barmaid (she was of considerable bulk) sank from view with a desperate slowness behind the counter and I flung myself tight up against the other side, my taxi driver beside me. He still had his glass in his hand and the beer shot across the floor, making a dark stain and setting the sawdust afloat. The soldier too had made for the bar counter and wedged the girl on his inside. One of her shoes had nearly come off. It was an inch from my nose; she had a ladder in her stocking.

"My hands were tight-pressed over my ears but the detonation deafened me. The floor rose up and smashed against my face, the swing door tore off its hinges and crashed over a table, glass splinters flew across the room, and behind the bar every bottle in the place seemed to be breaking. The lights went out, but there was no darkness. An orange glow from across the street shone through the wall and threw everything into a strong relief.

"I scrambled unsteadily to my feet and was leaning over the bar to see what happened to the unfortunate barmaid when a voice said, 'Anyone hurt?' and there was an AFS man shining a torch. At that everyone began to move, but slowly and reluctantly as though coming out of a dream. The girl stood white and shaken in a corner, her arm about her companion, but she was unhurt and had stopped talking. Only the barmaid failed to get up." —from *The Last Enemy* by Richard Hillary

As Hermann Goering met with some of his fighter pilots in his private railway car at Cap Blanc Nez near Wissant after the raid of the 7th, he was plainly angry. The teleprinter spat out the first reports of the day's action and his rage grew. Of the first wave of his bombers dispatched, nearly forty of 247 had been lost. What he was unable or unwilling to accept was that, having put up 600 fighters in escort of his bombers, that huge mass of aircraft had created chaos in the London sky, making the protection work of his fighter pilots all but impossible. "Your job is to protect the bombers" he railed. "Don't tell me the sky is full of enemies—I know they haven't more than seventy fighters left."

In the aftermath of the great raid on London, Assistant Divisional Geoffrey Blackstone of the London Fire Brigade felt the weight of his own responsibilities. Of the 30,000 Brigade members, 28,000 were wartime auxiliaries, many of them conscripts, but even more were volunteers. What ninety percent of them had in common was never having fought a fire of any kind. In peacetime a major fire required some thirty pumps, as fire engines were known. By midnight on 7 September, there were nine separate 100-pump fires burning across Dockland, the Woolwich Arsenal, and the Bishopsgate Goods Yard. At Rotherhithe, on the river Thames, Geoffrey Blackstone arrived at the Surrey Commercial Docks. Blackstone was tall and fit, an ex-public school boy, who had raced over from the tennis party he had been attending at a friend's in Dulwich, South London, to take charge of a huge fire of resinous timber stacked twenty feet high over some 250 acres. Even the dockland roadway was ablaze. At Paget's Wharf fire station, the senior officer was on the phone with Fire Brigade headquarters in Lambeth: "Send every pump you've got. The whole bloody world's on fire."

So intense was the heat from this blaze that the paint was blistering on the fire boats moving past the scene some 300 yards away near the opposite shore of the river. Enormous solid embers were being tossed like cabers over into distant roads to start fresh fires. Telegraph poles were bursting into flame, as were fences. Burning barrels of rum were exploding like liquid oxygen cylinders. Choking, nox-

above: Mothers and children awaiting trains from London in the Blitz period, 1940-41.

ious fumes were rising from the many rubber fires, fires started by the hundreds of high-explosive bombs and cannisters of incendiaries from the enemy bombers. And now armies of terrified rats ran through the streets of this burning neighbourhood. Brigade fireman William Ward: "I don't think any fireman has ever seen anything like it before."

The next day, Sunday 8 September, the war correspondent Ed Murrow told his thirty million American listeners, from Studio B4, Broadcasting House: "This ... is London." He recalled the previous night: "The fire up the river had turned the moon blood-red. Huge pear-shaped bursts of flame would rise up into the smoke and disappear. The world was upside down."

"When people's ill, they come to I / I physics, bleeds, and sweats 'em; / Sometimes they live, sometimes they die. / What's that to I? / I lets 'em." —*On Himself* by Dr J.C. Lettsom

"What is moral is what you feel good after and what is immoral is what you feel bad after." —from *Death in the Afternoon* by Ernest Hemingway

The Defence

"I have spoken to your uncle at length about your desire to be a pilot and he has advised me against it. Hundreds of pilots are chasing a handful of jobs. Some of the best and most experienced have applied to join his firm. So what makes you think you'll be any better than they are and can walk straight into a job?"

Geoffrey Page had always wanted to be a fighter pilot. But those discouraging words from his father when Geoffrey was a very young man had been painful. He tried to explain that what he really wanted in life was a permanent commission in the Royal Air Force and an air force career. His father, though, was adamant. "Your uncle [Frederick Handley Page, aviation pioneer and founder of Handley Page Limited] is prepared to find you a place in the Company if you qualify as an engineer. And that is precisely what I would do if I were in your shoes. Of course, if you insist on going to Cranwell, you'll have to pay for it yourself, as I don't intend to provide the money for such stupidity."

So, Geoffrey had to bite the bullet and accede to his father's wish for him. He reluctantly entered London University to study engineering. It was sometime later that he discovered the real reason for his father and uncle having discouraged his interest in flying. Their younger brother had been a pilot and was killed in a flying accident.

Geoffrey learned many things in London, not least that if he were able to pass the strict medical examination, he could enrol in the University Air Squadron and have the opportunity to learn to fly with free RAF training. He jumped at the chance and took every bit of what spare time he had to take flying lessons at London's Northolt airfield. The intent of the University Air Squadron concept was to both encourage undergraduates to choose a Royal Air Force career, and to create a reserve of partially-trained officer pilots who could be quickly brought up to operational standards in the event of a war.

Quite soon it was clear that, while he was becoming a competent pilot, his academic progress was less impressive and in the summer of 1939 he received an ultamatum from his parents—either continue his engineering studies without the distraction of flying or leave university to make his own way in the world. Before he was required to make that decision, Hitler came to his rescue. The German invasion of Poland caused Geoffrey and a great many others to be called up for flight training at RAF Cranwell.

Like many people who make a life for themselves in flying, Officer Cadet Page was something of a romantic. He had long since read about and greatly admired the Royal Flying Corps pilot, Captain Albert Ball, of First World War fame. A hero and inspiration to Geoffrey, Ball had been known for hunting alone deep behind enemy lines. He had accounted for forty-seven enemy aircraft shot down and had been awarded the Victoria Cross for his achievements.

Geoffrey was greatly motivated to do well in his air force training. He performed with distinction at Cranwell and did especially well in his advanced flying training—so well that on graduation he was, against his wishes, assigned as a flight instructor. When he expressed his disappointment to the chief flying instructor he was told: "You must remember that to be sent to central flying school to be trained as an instructor for future instructors is about the highest compliment the Air Force can pay a pilot. We didn't give you an exceptional assessment just to get you shot down! And another thing to re-

below: A German aerial reconnaissance targetting photo of central London made in 1939.

London

Karte 1:100000 1:6...
Blatt Bl
34 107...

aufnahme: Länge (westl. Greenw.): 0° 8′ 30″, Breite: 51° 31′ 0″ Nachträge:
R 51 Mißweisung: 10° 50′ (Mitte 1938) 4. 6. 39.

500 0 500 1000 m

Maßstab etwa 1: 15 500 (1cm ⊢ 155 m)

Wing Commander Alan
Geoffrey Page, of 56
Squadron, endured more
than forty plastic surgery
operations after being bad[ly]
burnt in the downing of h[is]
Hurricane in August 1940[.]
He was a founding memb[er]
of the Guinea Pig Club, th[e]
Second World War burn
patients of the surgeon Sir
Archibald McIndoe at the
Queen Victoria Cottage
Hospital, East Grinstead. R[e]
turning to operational flyi[ng]
he finished the war with
seventeen aerial victories.

member, you're not going to like this, but good pilots are often bad fighter pilots. A fighter pilot needs to be very ham-fisted on occasions and you're just not made that way. Sorry!"

But once again Mr Hitler came to the rescue. With the fall of France to the Germans in 1940, the Air Ministry posted Geoffrey to 66 Squadron, a Spitfire outfit based at Horsham St Faith, which is now the Norwich airport. At the squadron, when his new commander learned that Geoffrey had thus far flown nothing hotter than a Hawker Hart, he roared: "Christ! What will they be sending us next? For a start, we may as well establish that you're a damned nuisance. I think it might be wiser if I sent you away for a conversion course." But he eventually agreed to let Geoffrey have a go in a Spitfire.

"I had been sitting in the cockpit for half an hour memorizing the procedures for take-off, flight and landing. An airman climbed onto the wing behind me to help me with my parachute and harness. Word spread swiftly that an unusual first solo on type was taking place and soon ground and air crews were gathering to watch with morbid interest. As soon as I was properly strapped in, the squadron commander climbed onto the wing for a final word. 'Don't forget, taxi out quickly and turn her into the wind. Do a quick check and then get off. If you don't, the glycol will boil and so will my blood. Good luck!'

"I responded with a nervous smile, closed the tiny door and turned to face the mass of dials, buttons and levers. For a moment panic seized me and the temptation to undo the straps and get out was very great. The enquiring voice of the airman standing by the starter battery reminded me of the engine starting procedure, and my nervous feeling passed with the need for concentrated action. Throttle about half an inch open. Gas on. Nine full strokes in the KI-gas hand priming pump for a cold engine. Propeller in fine pitch. Brakes on. Stick held back. Press the starter button. I raised my thumb, the waiting airman replied with a similar sign, and I pressed the starter button firmly. The propeller began to rotate.

"A trickle of sweat ran down my forehead. Suddenly the powerful engine coughed loudly, blew a short stream of purpley-white smoke into a small cloud and roared into life. Remembering that I had little time to spare before the temperature reached the danger mark of 110 degrees, I waved my hands across my face. The waiting airman quickly ducked under the wing and pulled away the restraining chocks. Glancing down, I was alarmed to see that the glycol temperature had risen from 0 to 70 degrees. Releasing the brake, I eased the throttle open and the surge of power carried the aircraft forward rapidly over the grass.

"Was everything ready for a quick take-off? I figured I'd better call up Flying Control and get permission to take off immediately. Pushing over the switch on the VHF box, I tried to transmit. 'Idiot!' I said to myself, switch the damned thing on. Another glance at the temperature showed 95 degrees and still a long way to go before turning into the wind. The radio came to life with a whine. The controller's voice was soothing and for the first time since strapping into the narrow cockpit, I relaxed slightly.

"The temperature now read 105 degrees and there were still a few yards to go, plus the final check. Softly I prayed for help. Temperature 107 degrees. Now for the drill: R-A-F-T-P-R. The radiator—God alone knows how many times I'd vainly tried to open it beyond its normal point to try to keep the temperature down. Airscrew in fine pitch. Flaps—OK. Temperature—109 degrees. I abandoned the rest of the cockpit drill and, opening the throttle firmly, started the take-off run. Working the rudder hard with both feet to keep the sensitive little machine straight, I was too busy for other thoughts. Easing the stick forward, I was startled by the rapidity with which she responded to the elevator con-

trols. The long nose in front of me obscured the rapidly approaching end of the airport, but by looking out at an angle, I was able to get an idea of how far away it was. If the glycol boiled now at this critical stage . . . looking back into the cockpit again, I saw the hated instrument leering at me—110 degrees.

"Accompanying the feeling of fear was a new sound. The wheels had stopped drumming and a whistling note filled the air. The Spitfire soared gracefully into the air, thankful, as I was, to be away from her earthly bonds.

"Inside the cockpit I worked desperately to get the undercarriage raised. The CO had explained to me that because the starboard aircraft leg hung down in front of the radiator when the wheels were lowered, this affects the cooling effect of the airstream.

"The Spitfire was now about twenty feet up, gaining speed rapidly and skimming over the trees and hedges. I selected 'wheels up' and gave the handle a first stroke. The engine cut out for an instant and the nose plunged earthwards. Being unused to the technique of keeping my left hand absolutely still

far left: He 111s en route to London; centre top: Identification views of the Hawker Hurricane; above: Squadron Leader Caesar Hull of 43 Squadron, RAF, stationed at Tangmere in West Sussex during the Battle of Britain; left: Group Captain Peter Townsend, right, talking with his rigger and fitter in 1940.

left: Group Captain Douglas Bader flew Hurricanes with 242 Squadron at Duxford near Cambridge in 1940; right: Sergeant John Burgess of 222 Squadron at RAF Hornchurch in the Battle of Britain; far right: an RAF pilot in the cockpit of his Spitfire; below: A cannon-armed Spitfire taxiing from its station reventment; bottom right: Group Captain Brian King- come, in Spitfire, was acting commanding officer of No 92 Squadron at Biggin Hill in the Battle of Britain.

A publicity photo used on recruiting posters for the RAF in 1940; bottom
right: Hawker Hurricane fighters being refuelled in 1940.

while the right one moved forward, I had inadvertently pushed the control column forward simultaneously with the first pumping stroke, thus causing the machine to dip suddenly. The negative G placed on the carburettor had caused a temporary fuel stoppage. Some trees flashed by alongside the aircraft as a frightened pilot hauled back on the stick, and soon I was soaring skyward again, pumping frantically after removing my left hand from the control column. At this stage it was obvious that the Spitfire could handle herself better than I could. After this nightmare, the green light finally shone on the instrument panel, indicating that the wheels were in the locked up position, and the engine temperature gauge showed a healthy fall.

"Now I had some breathing space, so I was able to look about and concentrate on the other aspects of flying the aeroplane. Throttling back the engine and placing the propeller in coarse pitch, I allowed myself the luxury of relaxing slightly and looked down on the beauties of the Norfolk Broads. However, the pleasures of the English countryside didn't last long. Glancing down and behind me, I was horrified to discover that the airport was nowhere in sight. The swiftness of the Spitfire had soon taken me out of sight of the landing ground, and although homing facilities were available over the R/T, pride stopped me from calling the Flying Control tower for assistance. Ten minutes later, relief flooded through me when the unmistakable outlines of Norwich Cathedral appeared out of the summer haze, and from there the airport was easy to find. A minute later the graceful plane was banking round the circuit preparatory to landing.

"R-U-P-F—radiator, undercarriage, pitch, and flaps. This time the pumping down of the wheels came

above: German gunners over England; left: Severe bomb damage in London during the Blitz of 1940-41; above right: The balloon barrage over London with Buckingham Palace in the foreground.

quite simply, and the other essential procedures prior to the final touchdown followed. The exhaust crackled delightfully as the engine was throttled back and the plane came in gliding fast over the boundary hedge. I eased the stick back and the long nose rose up and cut out the forward view of the landing run. Looking out to the left, I carefully judged the height as the Spitfire floated a foot or two above the grass. As soon as the machine had come to a halt I raised the flaps and undid the tight-fitting oxygen mask. The pool of sweat that had collected trickled down my neck. With a newly born confidence, I taxied the machine back towards the waving airman near the hangar. The feeling of achievement obliterated the memory of the fear I'd felt during most of the flight, and now I felt justified in taking a place among my fellow fighter pilots."

Roger Hall was flying as Yellow Two this day. There was only a slight wind blowing from the coast to the south and the sky was cloudless and pale. The phone in dispersal jangled and Ops alerted the pilots to stand by for a scramble shortly. The call served to heighten the tension every one of them felt. Ops was watching a plot that showed a large force of aircraft building up near Cherbourg. Each of the pilots felt as if he were in the starting blocks for a race, waiting for the crack of the starter's pistol. Then, after what seemed an interminable wait, the phone went again and they were off at a run.

They were scrambled towards Southampton and told to climb to angels three-zero. After take-off the two separate flights formed up as a squadron in a wide V formation and as they passed through 10,000 feet the CO ordered them back into Flight formation.

As they reached 20,000 feet, the CO called control for an update. Roger noticed several RAF fighters coming up in the north and east, possibly as many ten squadrons, nearly all Hurricanes, but there was also another Spitfire outfit among them. For once, it seemed to him, the odds would likely be close to even.

The increasing altitude brought the start of vapour trails from their engines. They soon arrived at angels three-zero and the CO so advised control. Now another set of vapour trails had come into view well above the height of the Spitfires. They were Me 109s, at least fifty of them and were almost certainly the advance party for a bomber stream. Looking down and ahead, Hall spotted the first of the enemy bomber formation at about 15,000 feet. Some 2,000 feet below his squadron, he saw some squadrons of Hurricanes ready to jump the German bombers, which appeared to number 100 or more. The bombers were stepped up in ranks front to rear and were packed rather closely together in a mix of Heinkel 111s and Dornier Do 17s.

The CO gave instructions for the pilots to adjust their positions. Roger was number two to Chumley, who admonished him now to keep weaving, which Roger was doing with exaggerated enthusiasm as the tail-end Charlie of A Flight. Between the intense glare of the sun and the dazzling high-white of the many vapour trails, he was challenged to keep track of the hostile aircraft in the area. As always, he was wary of the predictable tendency the German pilots had for attacking from out of the glaring sun and was constantly alert to the tactic and straining to see the first sign of it. His ears were being bombarded with the barks and squeaks and erratic chatter of transmissions to and from squadron aircraft and controllers, all of which indicating that air combats were under way somewhere near, but he kept reminding himself not to be distracted. His entire concentration must remain focused on the threatening 109s that he could sense were about to appear.

As usual in such circumstances, he was sweating despite the relative cold of altitude and the sweat poured off his brow and off his oxygen mask as he strained through a series of tight serpentine manoeuvres. He squinted at the sun and was sure he was seeing some of the 109s start their attacking dive. He shouted into the R / T "Look out Maida aircraft—109s coming down Now!"

Roger ceased weaving and closed in tight behind Chumley as Yellow Three now did the same behind him. The three Spitfires entered a tight turning circle which was immediately pierced by eight, possibly nine Messerschmitt fighters flashing down, guns blazing. With considerable relief, Roger realized he had not been hit and he wondered if any of his fellow pilots had. With the passing of the 109s, two sections of the Spitfires rolled over onto their backs, twisting vertically down into a spiral, and twisted into an attacking line astern. Someone yelled "Keep a look out behind, Roger!" and he was aware of the probability that some enemy fighters would be following them, having decoyed the Spitfire pilots with a little ruse meant to tease them into a nice, ripe straight line. Roger had doubts about the tactic his leader had elected, but now they were committed and had to make the best of the situation.

The German fighters and the following Spitfires were heading straight down at maximum indicated airspeed. Roger chanced brief occasional glances in his rear view mirror, not really all that concerned that anyone could haul him in and overtake him at this speed. Wrong. White streaks of gunfire immediate overhauled his Spitfire and he simultaneously hit the transmit button and yelled "109s behind us, A Flight". He yanked the control column right to instantly pull away from the white streaks. The CO called "Split up into sections, A Flight, and shake 'em off!" as he pulled his own section into a screaming tight turn to port. Chumley was now leading Roger out of their dive and through an extremely tight, slightly climbing turn. In their dive, they had lost considerable height and now found themselves virtually level with the German bombers and in the midst of the main battle. The needle of Roger's

A WAAF barrage balloon detachment at work in Portsmouth, England, 1940.

altimeter was on 17,000 feet. Roger checked his six o'clock position to be sure no one was behind them. When he informed Chumley, he was told to keep weaving. Now the main battle was slightly below and to the north of them. As he watched, Roger saw several aircraft that had been hit and were trailing thick black smoke. A number of parachutes were descending, and Chumley said "We'd better get into the bombers, Roger", pointing them toward the fight ahead.

"Christ, I'm on fire!" Chumley yelled. There was smoke sweeping past Roger's plane and he saw the propeller blades of Chumley's Spitfire begin to slow and knew that his leader was evidently throttling his engine back. Then the propeller stopped. Expecting Chumley's plane to burst into flames at any second, Roger pulled back a bit, and he could think of nothing to say to his friend. He watched Chumley smoothly pull the crippled Spitfire around in a shallow turn to the right, slowing it down. He expected Chumley to bale out and hoped he would do so quickly. He thought that Chumley's engine must be cooling some as the smoke had lessened a bit. "I'm going to make for Tangmere, Roger." And with that, Chumley began descending in the general direction of the Sussex airfield to the east of Southampton Water. Roger asked him if he was OK and Chumley said "Oil pipe's gone for a Burton." Roger said he would stick around until Chumley got down and Chumley replied "OK. I'll be all right, I think."

As they passed through 5,000 feet and approached the field at Tangmere, the smoke from Chumley's engine had nearly stopped and he was setting up for a landing. Roger remained at 5,000, watching as Chumley managed to bring the Spitfire in for a safe, wheels-down landing and get out of the plane as soon as it came to a stop. Smoke was again pouring from the fighter and crash tenders and fire engines were already at the scene spraying it with foam.

Hurricane pilots are scrambled during the Battle of Britain.

Foreign Aid

They also fought for Britain, coming as they did from Poland, Czechoslovakia, Norway, New Zealand, Australia, the United States, South Africa, Belgium, Canada, Free France, Ireland, Jamaica, Palestine, Holland, Newfoundland, and Southern Rhodesia, many speaking little or no English, at least when they arrived and relatively little thereafter. But they shared the common wish to fight and defeat the German enemy and to do all they could to help the British keep that enemy out of England. They helped preserve the country as the primary base for the British and, eventually, the American bombers that would carry the war to German targets and cities in the great strategic round-the-clock bombing campaign. Many of these foreign airmen had left their homelands to escape Nazi occupation, most had left family and loved ones, often suffering considerable hardship in their struggles to get to England. None of them knew if or when they would ever return to their native countries and see those they had had to leave behind. But they all knew full well why they come to England.

New Zealander Al Deere had been the middle-weight boxing champion of the Royal Air Force and had the broken nose to prove it. Hailing from Wanganui. The tall, well-built Deere had a legendary appetite to match his enthusiasm and his flying ability, and was nearly always among the first at the breakfast table. He was truly one of the best of the best, and like his best friend, fellow New Zealander Colin Gray, since his teenage years he had no greater ambition than flying in the RAF. He had applied for a short-service commission and was among the first dozen to be accepted from the first 2,000 applicants.

As far as he was concerned, the Battle of Britain had begun with Dunkirk. For Al, the battle actually began on 23 May 1940. His squadron, No. 54, had returned to Hornchurch that morning after encountering no enemy aircraft on their dawn patrol flight. After a late breakfast, Deere was asked by Squadron Leader James Leathart to meet him and Johnny Allen out at their dispersal immediately to do a flight over to Calais-Marck on the French coast and pick up the CO of 74 Squadron who had been forced to land there with engine trouble. Leathart would be flying the two-seat Miles trainer. Al and Johnny would be escorting him in Spitfires. They intended to cross the Channel at sea level and hopefully avoid German fighters. When he landed at Calais-Marck, he would keep the engine running while the two Spits orbited the field. The trip across was uneventful and when they arrived over Calais, Al told Johnny to orbit the field at 8,000 feet and guard against their being bounced by enemy aircraft, while he would orbit low to protect the Miles. But when Johnny got to 8,000 feet he shouted over the R / T: "Al, they're here. About a dozen Huns below me and heading towards the airfield. I'm going in to have a go at them." Al said he would try to warn Leathart not to take off.

The Miles had no R / T. All Al could do to warn Leathart was to waggle his wings and hope the Squadron Leader got the message. He didn't, and began his take-off run in the Miles. A Bf 109 fighter appeared over the runway and opened fire on the training plane as Al chased the German into a turn. Just then, Johnny called him to say that he was surrounded and needed help. Al told him he would be right up to help as soon as he killed the bastard in front of him, which he then did when the German pulled up and gave him a near perfect shot. The burning 109 fell onto the beach. Al then noticed to his relief that the Miles appeared safely parked near the airfield perimeter fence.

As Al climbed to come to the aid of Johnny, he encountered two more 109s. He manoeuvred into

left: New Zealanders Colim Gray and Al Deere came to England to fly with the RAF in the Second World War; below: A Polish fighter pilot, one of many who chose to fly with the RAF, on the wing of his Spitfire at RAF Northolt in 1942.

firing position behind one of the German fighters, put a long burst into it and watched bits come off as it fell to earth. Al then chased the other fighter but found that his ammunition was exhausted when he tried to fire. The German then ran for his base.

Johnny called Al to say "I'm just crossing north of Calais but am rather worried about my aircraft. I can't see any holes but felt hits and she doesn't seem to be flying quite right. I'll make for the North Foreland at my present height of 8,000 feet. See if you can join up." When they met up, Al inspected the other Spitfire and decided that Johnny could probably make it back to Hornchurch, where they both later arrived safely. Leathart and his passenger arrived moments later in the Miles and told them how they had watched the overhead air battle from the comfort of a ditch near the Calais-Marck airfield boundary. Leathart: "When I left the ground I saw from Red One that something was amiss. Almost at once a 109 came down and began firing at me. I pulled around in a tight turn and saw the Messerschmitt shoot past me. I banged the aircraft on the ground and left the cockpit quickly, my passenger and I diving into the safety of the ditch which ran along the airfield perimeter. Then we saw a 109 come rushing out of the clouds to crash in a huge explosion a few hundred yards away. Right after that another 109 blew up as it hit the sea to our left. From the comparative safety of the ditch, we caught brief glimpses of the dogfight as first a 109 and then a Spitfire came hurtling through the cloud banks and screamed upward again. It was all over in about ten minutes and, when it seemed safe, we made a hasty take-off and a frightened trip back to England."

Later in the Battle, Deere: "This was the time when 54 Squadron should have been moved out of the line for a rest. Without a squadron commander, only four survivors with any experience of leading a flight, and no other pilot really up to the standard required to lead a section, the squadron was in an extremely bad way. It was a grim situation not made any easier to bear by the move out, on that very day, of 65 Squadron to a northern airfield for a rest. Once again 54 Squadron was to act as guinea pig for the blooding of the new squadrons being brought into the Battle. In the twenty-one days since our return, I had not been off the airfield, except for one unhappy night in hospital, nor had any of the four leaders on whom the squadron now depended. Indeed, the mess by daylight was an unfamiliar sight—when dawn broke we were already at dispersal, and it was after dark when we returned to the mess. The strain had almost reached breaking point. The usually good-natured George was quiet and irritable; Colin, by nature thin-faced, was noticeably more hollow-cheeked; Desmond, inclined to be weighty, was reduced to manageable proportions; and I, although I had no way of knowing how I appeared to the others, was all on edge and practically jumped out of my skin when someone shouted unexpectedly over the R/T. But still we continued to operate—there was no alternative.

"In the late afternoon of August 28th I was airborne at the head of the squadron for the third time that day; the now familiar gaggles of 109s were our inevitable playmates. In the melée that followed the first attack, I was once again on the receiving end—but this time of a Spitfire's guns. I had fastened on to the tail of a Me 109, one of three in close line astern formations, and was trying to close in to firing range, when a Spitfire dived in from above and pulled around behind me. I clearly saw the RAF roundels as, fully banked in a steep turn, the aircraft was silhouetted against a blue sky. 'Good, he's coming in to give me a hand,' I muttered into my mask.

"Imagine my surprise to find that I, not the three 109s ahead of me, was the subject of his wrath. Before I could do anything about it, he had found his aim and I was riddled with a burst of fire which struck the fuselage and port wing, cutting my rudder control cables and seriously damaging the port elevator. One burst was all he allowed himself before breaking down and away underneath me. It was all over in a matter of seconds, but even in that time the 109s had made good their escape, no doubt encouraged by the support afforded them!

"I throttled back immediately and, by so doing, was able to keep going straight and level, but at a very reduced speed. There was practically no elevator movement and, of course, I had no rudder control. What was I to do? Stick with the aircraft and try a landing or do the sensible thing and bale out. I decided on the latter course, influenced by the fact that my radio was dead, and I was above a considerable amount of cloud and therefore unable to position myself in relation to an airfield. Also, there was no way of really knowing how serious was the damage to my fuselage and whether or not I could maintain elevator control at the lower and more turbulent altitudes. Any misjudgement on the approach requiring the use of throttle would certainly make the aircraft uncontrollable.

"There was to be no rolling onto my back this time. I made up my mind to take a header over the side in the conventional way, having trimmed the aircraft as best I could to maintain an even course.

"Thus it was that I found myself swaying in my parachute at 10,000 feet over Kent, with my abandoned Spitfire diving away to destruction somewhere below me, and my thoughts wandering back over the years to my childhood.

"There was a feeling of splendid isolation as, snugly held in my parachute harness, I swayed gently to and fro high above the cultivated acres of Kent which stretched out below me, here and there flooded with warmth as the slanting sun rays pierced the broken cloud from whose vaporous interior I had just emerged. To one side I could see the long straight stretch of highway from Canterbury to Gillingham— the Watling Street of earlier happier days—and on the opposite side a dark broken line marked the edge of the North Downs along whose edge the early Pilgrims had trudged their devout way to worship at Canterbury. In the distance, Detling airfield was plainly visible, a cleared space in the wooded area which sat astride the roadway that ran from Maidstone to Sheerness.

"The descent from 10,000 feet took about fifteen minutes and thus there was plenty of time to ruminate on the past. Also it afforded me an opportunity to practise side-slipping in my parachute. It was as well that I practised this manoeuvre for it enabled me to avoid a farm-house towards which I was blown on the last few feet of my descent. I missed it by inches and landed instead astraddle a plum tree in an adjoining orchard. Apart from a few scratches in unpleasant places, I was unhurt.

" 'Stay where you are, I've got you covered,' an angry voice reached my ears. Startled, I peered through the plum-loaded branches and straight down the barrel of a shotgun held in a business-like manner by an irate farmer who occupied the key position at the foot of the tree. In amazement and not a little afraid, I gaped at him. Finding my tongue, I said in most precise English, 'I'm British,' hoping that he would not misinterpret my New Zealand accent as belonging to a German trying to pretend he was English. Fortunately he didn't. 'Oh, I thought you might be a German, and I wasn't going to take any chances. Did you have to choose my prize plum tree in which to land? I was saving that crop.'

"In the course of extricating myself and my spread-eagled silk canopy from the tree, a great deal more of his prize crop fell to earth, much to his annoyance. I feel sure that had I been a German, he would have let me have both barrels. As it was, he grudgingly allowed me to use his telephone to contact Detling airfield from which a car was sent out to pick me up."

The twenty-three Polish pilots of 303 (Kosciuscko) Squadron flying from RAF Northolt, were wholly contemptuous of the assignment they had been given on Friday afternoon, 30 August 1940. At 10,000 feet, north of the city of St Albans, in Hertfordshire, their Hurricanes were to rendezvous with six

Blenheim light bombers to execute dummy attacks on them. Having all been trained for two years with more than 500 flying hours each, and having come to England and the RAF specifically to kill Germans, they took a dim view of the exercise.

As one of them, F/O Ludwig Paszkiewicz, watched, a Hurricane went plunging downward towards a cluster of smoking rooftops. Paszkiewicz looked up and, at about 1,000 feet above his squadron, saw a large dogfight under way between Me 109s and a number of British fighters. He called his joint commander, S/L Ronald Kellett, "Hullo, Apany Leader, bandits ten o'clock!" There was no response, so Paszkiewicz went to full boost and roared up into the fight. The spirited Poles had nearly everything in their favour except for command of the English language, and because of that, they had not yet been granted fully operational status in Fighter Command. In fact, Kellett had indeed heard the warning from Paz, and had muttered a reply, "If you want to be a hero, be one." The Pole climbed and managed to get one of the slim German bombers, a Dornier Do 17, in his gunsight. He fired until the starboard engine of the bomber burst into flame and one of the German crewmen baled out before the plane dived to earth. That evening the impressed Kellett telephoned Headquarters, Fighter Command at Stanmore, to say, "Under the circumstances, I do think we might call them operational" and the next day 303 Squadron was duly reclassified. And after 31 August, Kellett was the first to admit that 303 Squadron was doing the work of two squadrons, in a month in which pilot wastage was nearing 120 men a week. In six weeks of non-stop combat, 303 destroyed 126 German aircraft for the loss of just eight of their own pilots. The war correspondent Dorothy Thompson, writing in the New York Times: "The Poles are pure courage."

Second from left, Group Captain Johnny Kent, a Canadian, led the Polish pilots of 303 Squadron at Northolt in 1940.

Thirty-year-old Adolph Malan was the stand-out of the twenty-three South African pilots who flew in the Battle of Britain. Malan had served for some time before the war aboard the ships of the Union Castle Steamship Line and was ever after known as 'Sailor.' Many came to believe, with some justification, that he was greatest RAF fighter pilot of the war, possibly the greatest ever. Sailor discovered that the four Browning machine-guns of a Spitfire, firing 1,260 rounds a minute, had the firepower equivalent of a five-ton truck hitting a brick wall at sixty miles an hour.

On trips into London, both Sailor and Al Deere tended to leave their DFC medals back at the base, so modest were they. On one occasion, Sailor told the war correspondent Quentin Reynolds that he had never in his whole life read a book. "You see, I have no imagination. If I had any imagination, I'd have been dead by now." Yet he produced one of the most cogent documents to come out of the Battle of Britain and the entire war—Ten of My Rules for Air Fighting, a poster of which had pride of place in most dispersals of Fighter Command and of the American Air Force the world over. It was the Ten Commandments that guided, and undoubtedly saved the lives of many British and American fighter pilots in the Second World War.

Sailor was a superb tactician and a great fighter leader with good, basic values who spent extra time with the newest, least experienced pilots in his command, showing them the ropes and doing all he could to keep them from making the kinds of mistakes that might get them killed. He was credited with downing thirty-five German planes in the war. At war's end, he returned to his native South Africa where he helped organize a quarter of a million war veterans and others into the Torch Commando, a campaign to try to change the racist policies of the country.

South African Group Captain Adolph G. 'Sailor' Malan, of 74 Squadron, led the Biggin Hill Wing and has been acknowledged by many as the greatest fighter pilot in the history of the Royal Air Force. His Ten Rules of Air Firghting probably saved the lives of hundreds of Allied fighter pilots in the Second World War and since. He died in 1963 of Parkinson's Disease, aged fifty-three.

Fifty years after he flew with 308 (Polish) Squadron in the Battle of Britain, Kazimierz Budzik was still the same keen-eyed, eager, smiling fighter pilot-type he had been in the summer of 1940, more than willing and ready to talk about his experiences in that bright blue time. Kaz had been commissioned a Second Lieutenant in the Polish Air Force just before he had to flee the country and the Nazi invasion. He first went to France where he flew Dewoitine and Potez fighters from the Pol air base near the Pyrenees until the fall of the country in June. With French help he was able to reach Casablanca and then Gibraltar from which he was finally able to sail to England. After reaching Liverpool, he entered RAF training and was then posted to 308 (Polish) Squadron on Spitfires. He came from Kraków originally, so it was most appropriate that he was made a part of 308, the City of Kraków Squadron, based at RAF Northolt. The primary mission of the squadron was bomber escort and low-level fighter sweeps over France.

For Budzik, service with the RAF was two operational tours of duty and a massive number of dog-fights, strafing attacks, night-fighter patrols, dive-bombing missions, and bomber escort missions. He achieved at least twenty kills and probables, but it was the doing itself, not the credit for the doing, that really mattered to him. His recollections of air combat with the squadron were expressed with candor, especially about their impatience and over-excitability: "We always, always opened fire much too early. The English seemed to have more patience and self-control, waiting until they got in close or manoeuvred into the best position before opening fire." The Poles flying in their 308 Squadron mostly used Polish in their brief R/T comments. "Most of us knew little English and, in any case, if in a tight corner or shouting a warning of danger, we'd be fumbling around for the right English words and someone would have soon been dead. You had to do it before you had thought about it."

One young Pole, P/O Jan Wiejski, posted to Budzik's squadron, simply didn't possess such lightning-like reactions, so Kaz told him: "Look here, why not see the CO about a transfer to another squadron? You just won't make it otherwise. Bombers, or Coastal Command, perhaps, is your thing." Wiejski knew Kaz was right, but he would not change; he'd see it through. "We were escorting bombers, but the operation just didn't go properly to plan. On the way back we saw a group of Fw 190s. I was one of the last in my section of six aircraft and, watching the 190s, we didn't see another group of them coming from behind. Suddenly, something in my mirror. A flash. Instinctively, I swerved and pulled around tightly toward our attackers and was head-on to about ten, maybe twenty, 190s. The next few seconds were spent just trying to survive. I did, but the rest of my flight—five aircraft in all—were shot down. Among them was Jan Wiejski. It was his first operational trip. He didn't stand a chance."

In the D-Day invasion on the Normandy coast of France, 308 Squadron was assigned as an element of the 2nd Tactical Air Force and was deployed in the ground-attack / support role. They were now flying Spitfire XIVs equipped with cannon and bombs and there was considerable satisfaction for the pilots knowing they were delivering real damage to the enemy. "I flew what must have been dozens of flights during the struggle for the Falaise Pocket. I don't mind admitting to a feeling of glee and excitement because, on the ground, I could see Jerries by the score looking up at me . . . terrified. Some of them had their hands above their heads as I roared over them. I remember thinking, 'Now you bastards. How do you like it?' It was a different story in 1939, though. They showed Poland no mercy then, but here they were now, begging for it. It was a wonderful experience."

Around the time of the invasion, 308 Squadron's Spitfires were being equipped with a new gyro gunsight. The remarkable innovation greatly increased the accuracy of their shooting. It made automatic corrections for deflection and mostly eliminated human error—but not Polish impatience and excitement. "They fitted these sights, but they never trained us how to use them. Instead, we had to find out for ourselves. For air-to-ground work, no problem. You just put the dot on the target and simply hit it. You couldn't miss. Of course, by this time, intervention by German fighters was comparatively rare. We just got on with our lovely job of bombing and shooting things up on the ground. Then, one day, there was a shout of 'Focke Wulf' over the radio. 'I'll get him!' I called, diving after him absolutely flat out. He was in my sights. Rrrrrrrmmmp . . . nothing. Not a single hit. Problem was that these wonderful new gyro sights needed a second or two to settle. I hadn't allowed for that. My impatience and excitement wouldn't let me. But now the chase was on. Down on the deck we were streaking for Germany. I wouldn't let him get there. No. Was that smoke or exhaust fumes I could see trailing out behind? Well, it soon would be smoke. In my sights . . . gyro settled now . . . fire! Nothing. All my ammo had gone in that first burst that had missed. But at least I had learned about gyro gunsights.

"After the invasion, we had been flying from Ghent in Belgium and were tasked to carry out a dive-bombing attack on Walcheren Island off the Dutch coast. As we approached the target a terrific amount of flak came up from all directions. When that happened, we learned that we could tease the gunners by holding off just out of range of their guns; then we'd judge the moment to dive when the

below: Dornier Do 17 medium bombers over England in the summer of 1940.

fire had slackened off, or maybe we'd fool the gunners into thinking that we were going away. Often this ploy worked quite well. But on this occasion I remember I was trembling with fear. Everyone experiences fear, but for me the fear went once we had committed ourselves to the attack—although the trembling soon returned. I was leading my section in when suddenly wham! I quickly turned the aircraft around and, despite the damage, I managed to get it down in a crash-landing. Once I was down, everything seemed so quiet, and yet, as I got out of the cockpit there was the sound of a ground battle all around me—heavy gunfire, rifles, and machine-guns. I quickly got myself to a ditch and thought that this was not a good situation for a fighter pilot to be in. Then some civilians turned up. From a distance I shouted to them, 'Go away!' and waved my pistol to reinforce my request. I was worried that they might turn me over to the Germans. But then a woman among them shouted in English, 'Are you British?' 'No, Polish.' Back came the reply, 'Down the road. We're waiting for them.' Once identities and allegiances had been established the civilians came closer and a man with a bicycle approached and said he'd take me to the British lines. I hopped on the bike and went off with him, but before I did, I produced my pistol again and told him I'd shoot him if he was taking me to the Germans. He didn't, so I didn't have to shoot him after all, and gave him cigarettes instead of lead."

below: Bomb damage in London during the 1940-41 Blitz night bombing attacks by the German Air Force.

The Raiders

Bomber and fighter squadrons of the Luftwaffe were relocating from their German bases to airfields near the French coast with the fall of France in June 1940. Their new fields put them within easy striking distance of their targets in England. For the fighter escorts, though, the Messerschmitt Me and Bf 109s that were charged with shepherding the Heinkel He 111s, Junkers Ju 88s, Dornier DO 17 bombers, and the Junkers Ju 87 dive-bombers, it was not much of an improvement. The German fighters were not provided with additional external fuel tanks and their range was severely limited. They were normally stretching it to the limit to make the round trip from a French base to an English target, with roughly a ten-minute allowance over England.

The German "bases" in France were probably best described as rough and ready, their French constructors having put much higher priorities on the airfields they had built in eastern France, guarding the French border with Germany. Further construction was carried out on the western airfields to bring them up to an acceptable standard for Luftwaffe operations, much of the work being done by French forced labourers euphemistically referred to by the Germans as special workers. During their early days on the crude French airfields in the west, the Germans spent much of their time on leave, enjoying the beaches along the Channel coast and the allure of the seaside towns and villages.

But with the issuance of Hitler's order of 16 July in preparation for Operation Sea Lion, the top-secret plan for the invasion of Britain that autumn, the holiday came to an abrupt end for the pilots and aircrews of the German air force in France. The Luftwaffe was equipped on the French fields with some 2,600 aircraft: 1,480 bombers and 1,120 fighters. Like their Allied airmen counterparts, the German fliers were keenly aware that every mission they flew across the Channel could well be their last. Because of that threat they tended to live life to the fullest between the missions. While there few comforts, luxuries, or even simple pleasures available to them in that circumstance, the Luftwaffe did make an effort to do what could be done towards maintaining the morale and spirit of the aircrews. In one such example, a Junkers Ju 52 transport plane was pressed into service making frequent flights over to the Channel island of Guernsey, which had been occupied by the Germans. There it was loaded with fresh fruit and vegetables, whiskey and cigarettes to be distributed on one of the French air bases in a welcome break from the canned food to which the fliers were accustomed. Ample diversion too was provided by the uniformed German *Helferinnen vom Dienst*, the female helpers of the service, many of whom contributed generously to the morale of the airmen there.

On the forward airfields in France, where possible, the pilots were generally billeted in barracks or commandeered civilian accommodation, while in the primitive conditions of these bases, the mechanics and ground personnel lived in tents on the airfields. Ever-present mud made their existence challenging at best and downright miserable much of the time. The fighter pilots of the Luftwaffe, the fortunate ones among them, were often housed in the comfort of requisitioned chateaux. When they returned from raids on England, they were well looked after for the most part.

Fall Gelb (Plan Yellow) was the German plan for an offensive against Holland, Belgium, France, and ultimately, Britain, and within it, the task of the German forces was to overrun and destroy as many French and Allied forces as possible to pave the way for a naval and aerial assault on Britain. The planning of Fall Gelb was the responsibility of Generaloberst Erich von Manstein and the staff of the Ger-

Heinkel bombers cruising low over the English Channel to avoid detection by RAF radar in the autumn of 1940.

man High Command. The extremely ambitious offensive was heavily dependent on powerful, effective air support of the infantry, armour and other ground forces on a spectacular, unprecedented scale. The equipment assembled by the German air force for Fall Gelb included 1,120 bombers, of which more than half were Heinkel He 111s, with the remainder mainly Dornier Do 17 light bombers. Additionally, there were 380 Ju 87 Stuka dive-bombers, all supported by 860 Messerschmitt Bf 109E fighters. That huge force, augmented by some 640 Messerschmitt Bf 110 *Zerstörer* twin-engine heavy fighters, 475 Junkers Ju 52 transports for supply delivery and paratroop operations, as well as 230 gliders, launched in surprise attack operations in the pre-dawn hours of 10 May 1940.

The initial actions against airfields in Belgium and Holland were followed by massive airborne landings and parachute drops with varying degrees of success achieved as the German forces met some strong resistance. They experienced particularly high losses among the fleet of Ju 52 transports, losses approaching forty percent. By 13 May, however, the Luftwaffe had secured air superiority over the entire length of the front. The Stukas used in this overall action were especially effective and successful under the conditions of the air superiority. The Bf 109 fighters greatly outnumbered and outperformed the French and British defending fighters, which were never really a serious deterrent to the tactical operations of the German air force. By nightfall of 26 May, German forces had accomplished much of what they had set out to do and the evacuation of the enormous British Expeditionary Force from the beaches of Dunkirk was under way.

In the skies over Dunkirk, the pilots of the German fighters experienced their first contact with the impressive Spitfire Mk I of the Royal Air Force. Many would have been alarmed—some of them shocked—to find that their 109s were being out-turned and out-climbed by the fast and agile Spitfires. Even flown at the absolute limits of their performance in the action, the 109s were, in the main, out-classed by the British fighter. Even though the Spitfire and Hurricane pilots had the distinct disadvantage of having to operate far from their home bases in England, the German fighter pilots were now being involved in many dogfights over the beaches and were frequently missing their planned rendezvous to escort their bombers, which led to heavier losses than anticipated for the bombers and for the Stukas over Dunkirk. It was the first occasion, since the start of the heretofore brilliantly successful blitzkrieg (lightning war) campaign of the Germans, in which they were required to fight an opponent of at least equal capability. As a result, the Luftwaffe was unable to prevent the British evacuation of the Allied troops at Dunkirk, an operation completed in the early hours of 4 June.

The German campaign for France was over by 25 June. They had occupied Paris on the 14th and, after the 25th, the Luftwaffe was rested and allowed time for a re-fit, to be ready for the next and key phase of the overall western offensive, the invasion of Britain. In the air over Dunkirk, Goering's air force had had exposure to the sort of powerful, determined opposition they would face when they would try to secure air superiority over the British Isles.

The German plans called for an invasion of Britain within three months of defeating the French, in order to accomplish it before the worsening weather of the autumn. There was general agreement among the planners that such a land assault was far preferable to the months, or possibly years, that might be required for a lengthy campaign of economic strangulation. While Hitler continued to look for a diplomatic breakthrough with the British, it was not forthcoming and, on 2 July he ordered that the preparations for Sea Lion go forward. The prerequisite for the cross-Channel invasion, however, was not merely air superiority now, but air supremacy, which the Luftwaffe must secure and maintain. They had to neutralize the aerial opposition by the RAF and destroy its ground facilities.

Throughout July, the Luftwaffe consolidated its assets in France, deploying its various dedicated units from the German-French border, across the country to the Channel coast. In the build-up for the offensive on Britain, 1,200 bombers, He 111s, Do 17s, and Ju 88s, 280 Ju 87 dive-bombers, 760 Bf 109E fighters, 220 Bf 110 fighters, fifty long-range reconnaissance and ninety short-range reconnaissance aircraft were assembled and made ready, for a total of 2,420 aircraft committed to the assault. Additionally, Luftflotte 5 maintained a reserve force of 130 He 111 and Ju 88 bombers, thirty long-range Bf 110 fighters, and thirty long-range reconnaissance aircraft for back-up and possible use as a diversionary force.

In the interim, before the main onslaught by the Germans, RAF Fighter Command used the precious time to re-equip its vital front-line squadrons, as well as to re-build its own reserves, all of which were sorely depleted following the Dunkirk operations. Over Dunkirk, Fighter Command lost more than 100 fighters and eighty pilots, killed or missing in action. These losses, though hard to bear, were relatively minor compared to the losses the command had suffered between 10 May and 20 June, the high points of action thus far. In that period, the RAF lost 944 aircraft in total, including 386 Hurricane fighters and sixty-seven Spitfires. By 5 June, the situation was so dire that only 331 Hurricanes and Spitfires, and a further thirty-six aircraft were available for operations. Thanks, though, largely to the efforts of the Minister for Aircraft Production, Lord Beaverbrook, the fighter aircraft inventory was soon growing impressively. The replacement rate for trained pilots and aircrew, however, was nowhere near

Heavy bomb damage inflicted on Portsmouth, England by
Luftwaffe bombers.

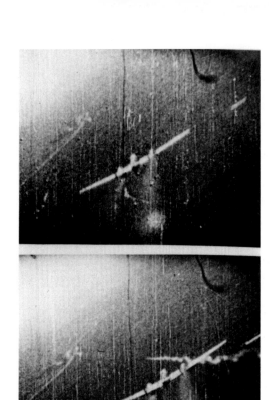

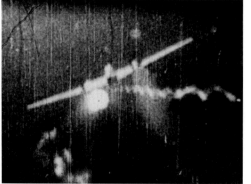

left: The final moments of a German bomber as captured by the gun camera of a Royal Air Force fighter in the Battle of Britain; above: The Spitfire of RAF Sergeant Pilot Denis Robinson, of 152 Squadron, flying from Warmwell. Robinson was shot down by a Me 109 while running in to land. "I was quite pleased with myself as the Spit slithered across the grass. Then suddenly, I felt her going up onto her nose and, I thought, over onto her back."

as impressive. There the shortages were crippling and seemingly insurmountable.

As for the aircraft, the great majority of the Fighter Command squadrons to that point were equipped with Hawker Hurricane 1s, an excellent aircraft of rugged construction, highly capable and a very stable gun platform greatly appreciated by its pilots. It took punishment well and was easily repaired, but was not as fast or as glamourous as the Spitfire and was generally considered inferior to the primary opposition, the Bf and Me 109 types. In the critical period of the Battle of Britain, the Hurricane would continue to be the mainstay of Fighter Command, however.

While at a distinct numerical disadvantage throughout the period before, and through the Battle itself, RAF Fighter Command had one significant advantage. Thanks to the progressive thinking of Air Chief Marshal Dowding and others in the command structure in the years immediately leading up to the Battle of Britain, the RAF had in place a wonderful Ground Controlled Interception (GCI) system, without equal anywhere. Utilizing a network of new radar installations situated around the English coastline, the incoming information on air raid threats was analyzed and passed along to the controlling stations and relayed by a Fighter Controller via radio-telephone to airborne squadrons. The system eliminated fuel-wasting standing patrols by the defending squadrons. This new, untested system held great promise and would prove itself efficient and extremely effective in the coming day and night offensive Battle.

In mid-July, the Luftwaffe reiterated the task at hand for its Luftflotten: the elimination of the Royal Air Force as a fighting force and a ground organization, and the destruction of the supply shipping to Britain in its ports. In two phases, the RAF was to be eliminated through attacking the fighter defences in the south of England, and to the north, by hitting RAF bases throughout England in a day and night bombing campaign against the British aircraft industry. This overall offensive was code-named *Adlerangriff* (Eagle Attack) and the official launch date was set as 10 August. The German planners estimated it would require four days to neutralize the RAF fighter defences, and four weeks more to destroy the entire RAF as a fighting organization. This accomplished, it was believed that Sea Lion could proceed in the first half of September.

The first job for Luftwaffe Fliegerkorps II, based in the Pas de Calais, and Fliegerkorps VIII, based in Normandy, was to gain air superiority over the Channel and over the British convoys. They would be supported by additional formations of fighters and Stukas. Henceforth German bombers were appearing in substantially larger formations over the Channel, the southeast coast and the Dover Straits in daylight, though they mostly confined their attacks to shipping, the ports and a few coastal airfields. And while the German bombers were so engaged, the bulk of the German fighter force in the west was focused on *Frei Jagd* (Free Chase) missions in an effort to entice the RAF fighters up for combat. The lure was at first irresistable for the squadrons of Fighter Command, but the lack of warning in that early stage, and the reliance on earlier tactics, placed the Spitfire and Hurricane pilots at a disadvantage, contributing to excessive losses. The RAF pilots soon learned a lesson and quickly adopted the German *Schwarm* and *Kette* tactics.

The Luftwaffe had to accept the fact that RAF Fighter Command was then still an effective fighting force and had not yet suffered the sort of losses the Germans had sought to inflict. For its part, Luftwaffe losses in the early stages had amounted to 192 combat aircraft destroyed with a further seventy-seven damaged. The Luftwaffe now had to attack targets further inland in the hope of destroying a greater proportion of the RAF fighter inventory.

A lesson for the Luftwaffe came in the discovery that the quality and even the quantity of German fighters being thrown into the effort was actually not sufficiently superior to that of the RAF. The skill, aggressiveness and determination of the RAF fighter pilots were such that the Germans seemed to need upwards of three times the number of fighter escorts they had anticipated to adequately support their bombers. Further, they soon found that the Bf 110C twin-engine fighter they depended upon for long-range escorted penetration missions was totally out-classed by the Spitfire and Hurricane in fighter combat. In such circumstances, the 110s were forced to form into defensive circles and, in extreme situations, needed the protection and support of the 109s. The strain of the Battle was telling on both sides and growing rapidly as the Germans intensified their bombing activity after 8 August. The capability of the RAF fighters was being stretched to the limit. For the Germans, they discovered in the early convoy attacks of this phase that the frightening, legendary Stuka dive-bombers were extremely vulnerable. Sixteen of the planes were lost on 18 August, establishing the plane as more of a liability than an asset. After that date the Stuka was withdrawn from the Battle.

Eagle Day was postponed by the Germans due to inclement weather lasting until the 13th. The opening of the main Luftwaffe offensive had to wait until the afternoon of that day. The massive bomber forces struck then at the ports of Southampton and Portsmouth, as well as the airfields of Detling, Eastchurch, and Middle Wallop, in a total of 1,485 sorties. They lost forty-six aircraft in the raids and were disconcerted by relatively poor bombing results at the start of their vaunted offensive. The next day the Germans redoubled their efforts to wipe out the RAF in the air, ruin its airfields and destroy its radar stations. And in the week that followed, the Luftwaffe attacks on the airfields of Fighter Command in southern England were ramped up greatly amid seemingly continuous air fighting over Hampshire, Sussex and Kent. The scope of its attacks on the aircraft industry and shipping was also increased. For their trouble, the Germans lost another 403 aircraft with 127 damaged, against RAF losses of 121 Hurricanes and fifty-four Spitfires.

24 August brought the third phase of the Battle of Britain. The RAF still existed despite the best efforts of the Luftwaffe to destroy it in the air and on the ground. Goering and Hitler were now becoming desperate. One factor making the Luftwaffe pay an especially high price for what they were trying to achieve was the absence of jettisonable fuel tanks on their otherwise very capable Bf 109E escort fighters. Their relatively short combat radius had been boosted slightly by relocating the majority of them to bases in the Pas de Calais, giving them greater penetration ability deeper into British airspace. The problem was compounded now by the ever-increasing losses being experienced by the German bombers. The German fighter leaders were being ordered by Goering to stay very close to the bombers they were supposed to be protecting, thus giving up nearly all the advantages designed into the 109. To help compensate for the reality of this sacrifice, many of the Luftwaffe's bombers and reconnaissance aircraft were being employed in diversionary feints; many fighters in complex sweeps, all manufactured in the hope of getting at the planes of Fighter Command while they were down on their fields being refuelled and rearmed.

The most punishing day of the Battle to date came for the RAF on 31 August when thirty-nine aircraft were destroyed and fourteen pilots killed. The personnel situation for Fighter Command looked better superficially than it actually was. The command had lost so many experienced Squadron and Flight commanders in the Battle so far that the bulk of the air fighting was being flown by a handful of remaining seasoned airmen, backed up by a gathering of fresh-faced, utterly inexperienced young pilots, most of whom would survive less than a week or two. Between 24 August and 6 September,

left: Messerschmitt Me 110s tucked into the trees on a forward airfield in France; below: A Heinkel He 111 crew member concentrates on filming over England.

Fighter Command lost 295 fighters and 171 damaged. Much more serious was the loss of 103 pilots killed or missing, with a further 128 wounded and lost to combat with injuries. The Luftwaffe losses for the period amounted to 378 aircraft destroyed and 115 damaged.

On the night of 25 August, a retaliatory attack against Berlin was launched by RAF Bomber Command. Hitler then used the attack on Berlin as his rationale for the revenge bombing of London, an offensive move he had until now rejected. On the apparent basis of the German successes in their bombing campaigns on Warsaw and Rotterdam, he seemed to think at this point that the British would react to such a campaign by suing for peace. On 2 September he had given Goering the order to attack the defences and population centres of Britain's largest cities in raids by day and night.

For the Luftwaffe leader it was an admission of defeat, though Goering still looked for something miraculous that would retrieve victory for him, something that would once and for all exhaust the British Fighter Command. But when Hitler turned his air force on London and the other big cities of Britain, doing so gave Fighter Command the breathing space it desperately needed then to regroup, rebuild and reinforce its ranks. Thereafter the pilots and crews of Fighter Command were able to operate largely with impunity in the air and on the ground.

As if further proof were needed that German air superiority over Britain, much less air supremacy, was as far from achievement as ever, the chaotic, astonishing air battle of 15 September over London certainly made the case. In that extreme series of encounters over the capital, some sixty German aircraft were shot down. It proved to be the death knell for Hitler's invasion plan, which he postponed indefinitely on the 17th. Still, he ordered his invasion fleet to remain at readiness until the 2nd October. In the fifteen weeks of the Battle of Britain the German air force lost 1,653 aircraft.

The Battle was essentially over. Now the attentions of the Luftwaffe were redirected primarily to the night bombing of London; the threat of invasion having been effectively neutralized by the Royal Air Force. As the weather gradually deteriorated from the glorious days of that summer, to the more typical rain and cold of England in autumn and winter, Goering shifted his bombing attacks on London to night raids. While refusing to admit defeat in the Battle of Britain, or even to refer to it as such, the German air force devoted itself to the goal of ultimately wearing down the resistance of the British people through constant bombing attacks on the population centres and by depriving them of vital supplies through continuing attacks on the major British ports and its shipping.

From 10 July through 31 October 1940, the German Luftwaffe suffered the loss of 1,733 aircraft against the loss of 915 RAF aircraft.

Churchill 1940

Winston Churchill, the British prime minister, rose in the House of Commons on 20 August 1940 and delivered what may have been the greatest speech of his illustrious career; a lusty, morale-raising, defeatism-defeating, pep talk of epic proportions and dramatic effect.

"Almost a year has passed since the war began, and it is natural for us, I think, to pause on our journey at this mile-stone and survey the dark, wide field. It is also useful to compare the first year of this second war against German aggression with its forerunner a quarter of a century ago. Although this was is in fact only a continuation of the last, very great differences in its character are apparent. In the last war millions of men fought by hurling enormous masses of steel at one another. 'Men and Shells' was the cry, and prodigious slaughter was the consequence. In this war nothing of this kind has yet appeared. It is a conflict of strategy, of organisation, of technical apparatus, of science, mechanics, and morale. The British casualties in the first twelve months of the Great War amounted to 365,000. In this war, I am thankful to say, British killed, wounded, prisoners and missing, including civilians, do not exceed 92,000, and of these a large proportion are alive as prisoners of war. Looking more widely around, one may say that throughout all Europe for one man killed or wounded in the first year perhaps five were killed or wounded in 1914-15.

"The slaughter is but a fraction, but the consquences to the belligerents have been even more deadly. We have seen great countries with powerful armies dashed out of coherent existence in a few weeks. We have seen the French Republic and the renowned French Army beaten into complete and total submission with less than the casualties which they suffered in any one of half-a-dozen of the battles of 1914-18. The entire body—it might almost seem at times the soul—of France has succombed to physical effects incomparably less terrible than those which were sustained with fortitude and undaunted will power twenty-five years ago. Although up to the present the loss of life has been mercifully diminished, the decisions reached in the course of the struggle are even more profound upon the fate of nations than anything that has ever happened since barbaric time. Moves are made upon the scientific and strategic boards, advantages are gained by mechanical means, as a result of which scores of millions of men become incapable of further resistance, and a fearful game of chess proceeds from check to mate by which the unhappy players seem to be inexorably bound.

"There is another more obvious difference from 1914. The whole of the warring nations are engaged, not only soldiers, but the entire population, men, women and children. The fronts are everywhere. The trenches are dug in the towns and streets. Every village is fortified. Every road is barred. The front line runs through the factories. The workmen are soldiers with different weapons but the same courage. These are great and distinctive changes from what many of us saw in the struggle of a quarter of a century ago. There seems to be every reason to believe that this new kind of war is well suited to the genius and the resources of the British nation and the British Empire and that, once we get properly equipped and properly started, a war of this kind will be more favourble to us than the sombre mass slaughters of the Somme and Passchendaele. If it is a case of the whole nation fighting and suffering together, that ought to suit us, because we are the most united of all the nations, because we entered the war upon the national will and with our eyes open, and because we have been nurtured in freedom and individual responsibility and are the products, not of totalitarian uniformity but of tolerance and variety. If all these qualities are turned, as they are being turned, to the arts of war, we may be able to show the enemy quite a lot of things that they have not thought of yet. Since the Germans drove the Jews out and lowered their technical standards, our science is definitely ahead of theirs. Our geographical position, the command of the sea, and the friendship of the United States enable us to draw resources from the whole world and to manufacture weapons of war of every kind, but especially of the superfine kind, on a scale hitherto practised only by Nazi Germany.

"Hitler is now sprawled over Europe. Our offensive springs are being slowly compressed, and we must resolutely and methodically prepare ourselves for the campaigns of 1941 and 1942. Two or three years are not a long time, even in our short, precarious lives. They are nothing in the history of the nation, and when we are doing the finest thing in the world, and have the honour to be the soul champion of the liberties of all Europe, we must not grudge these years or weary as we toil and struggle through them. It does not follow that our energies in future years will be exclusively confined to defending ourselves and our possessions. Many opportunities may lie open to amphibious power, and we must be ready to take advantage of them. One of the ways to bring this war to a speedy end is to convince the enemy, not by words but by deeds, that we have both the will and the means, not only to go on indefinitely but to strike heavy and unexpected blows. The road to victory may not be so long as we expect. But we have no right to count upon this. Be it long or short, rough or smooth, we mean to reach our journey's end.

"It is our intention to maintain and enforce a strict blockade not only of Germany but of Italy, France and all the other countries that have fallen into the German power. I read in the papers that Herr Hitler has also proclaimed a strict blockade of the British Islands. No one can complain of that. I remember the Kaiser doing that in the last war. What indeed would be a matter of general complaint would be if we were to prolong the agony of all Europe by allowing food to come in to nourish the Nazis and aid their war effort, or to allow food to go in to the subjugated people, which certainly would be pillaged off them by their Nazi conquerors.

"There have been many proposals, founded on the highest motives, that food should be allowed to pass the blockade for the relief of these populations. I regret that we must refuse these requests. The Nazis declare that they have created a new unified economy in Europe. They have repeatedly stated that they possess ample reserves of food and that they can feed their captive people. In a German broadcast of 27th June it was said that while Mr Hoover's plan for relieving France, Belgium and Holland deserved commendation, the German forces had already taken the necessary steps. We know that in Norway when the German troops went in, there were food supplies to last for a year. We know that Poland, though not a rich country, usually produces sufficient food for her people. Moreover, the other countries which Herr Hitler has invaded all held considerable stocks when the Germans entered and are themselves, in many cases, very substantial food producers. If all this food is not available now, it can only be because it has been removed to feed the people of Germany and to give them increased rations—for a change—during the last few months. At this season of the year and for some months to come, there is the least chance of scarcity as the harvest has just now been gathered in. The only agencies which can create famine in any part of Europe now and during the coming winter, will be German exactions or German failure to distribute the supplies which they command.

"There is another aspect. Many of the most valuable foods are essential to the manufacture of vital war material. Fats are used to make explosives. Potatoes make the alcohol for motor spirit. The plastic materials now so largely used in the construction of aircraft are made of milk. If the Germans used these commodities to help them to bomb our women and children, rather than to feed the populations and produce them, we may be sure that imported foods would go the same way, directly or indirectly, or be employed to relieve the enemy of the responsibilities he has so wantonly assumed. Let Hitler bear his responsibilities to the full and let the people of Europe who groan beneath his yoke aid in every way the coming of the day when that yoke will be broken. Meanwhile, we can and we will arrange in advance for the speedy entry of food into any part of the enslaved area, when this part has been wholly cleared of German forces, and has genuinely regained its freedom. We shall do our best to encourage the building up of reserves of food all over the world, so that there will always be held up before the eyes of the peoples of Europe, including—I say it deliberately—the Germans and Austrian people, the certainty that the shattering of the Nazi power will bring to them all immediate food, freedom and peace.

"Rather more than a quarter of a year has passed since the new Government came into power in this country. What a cataract of disaster has poured out upon us since then. The trustful Dutch overwhelmed; their beloved and respected Sovereign driven into exile; the peaceful city of Rotterdam the scene of a massacre as hideous and brutal as anything in the Thirty Years' war. Belgium invaded and beaten down; our own fine Expeditionary Force, which King Leopold called to his rescue, cut off and almost captured, escaping as it seemed only by a miracle and with the loss of all its equipment; our Ally, France, out; Italy in against us; all France in the power of the enemy, all its arsenals and vast masses of military material converted or convertible to the enemy's use; a puppet Government set up at Vichy which may at any moment be forced to become our foe; the whole Western seaboard of Europe from the North Cape to the Spanish frontier in German hands; all the ports, all the airfields on this immense front, employed against us as potential springboards of invasion. Moreover, the German air power, numerically so far outstripping ours, has been brought so close to our Island that what we used to dread greatly has come to pass and the hostile bombers not only reach our shores in a few minutes and from many directions, but can be escorted by their fighter aircraft. Why Sir, if we had been confronted at the beginning of May with such a prospect, it would have seemed incredible that at the end of a period of horror and disaster, or at this point in a period of horror and disaster, we should stand erect, sure of ourselves, masters of our fate and with the conviction of final victory burning unquenchable in our hearts. Few would have believed we could survive; none would have believed that we should today not only feel stronger but should actally be stronger than we have ever been before.

"Let us see what has happened on the other side of the scales. The British nation and the British Empire finding themselves alone, stood undismayed against disaster. No one flinched or wavered; nay, some who formerly thought of peace, now think only of war. Our people are united and resolved, as they have never been before. Death and ruin have become small things compared with the shame of defeat or failure in duty. We cannot tell what lies ahead. It may be that even greater ordeals lie before us. We shall face whatever is coming to us. We are sure of ourselves and of our cause and here then is the supreme fact which has emerged in these months of trial.

"Meanwhile, we have not only fortified our hearts but our Island. We have rearmed and rebuilt our Island. We have rearmed and rebuilt our armies in a degree which had been deemed impossible a few months ago. We have ferried across the Atlantic, in the month of July, thanks to our friends over there, an immense mass of munitions of all kinds, cannon, rifles, machine-guns, cartridges and shell, all safely landed without the loss of a gun or a round. The output of our own factories, working as they have never worked before, has poured forth to the troops. The whole British Army is at home. More than 2,000,000 determined men have rifles and bayonets in their hands tonight and three-quarters of them are in regular military formations. We have never had armies like this in our Island in time of war. The whole Island bristles against invaders, from the sea or from the air. As I explained to the House in the middle of June, the stronger our Army at home the larger, must the invading expedition be, and the larger the invading expedition, the less difficult will be the task of the navy in detecting its assembly and intercepting and destroying it on passage; and the greater also would be the difficulty of feeding and supplying the invaders if they ever landed, in the teeth of continuous naval and air attack on their communications. All this is classical and venerable doctrine. As in Nelson's day, the maxim holds, 'Our first line of defence is the enemy's ports.' Now air reconnaissance and photography have brought to an old principle a new and potent aid.

"Our Navy is far stronger than it was at the beginning of the war. The great flow of new construction set on foot at the outbreak, is now beginning to come in. We hope our friends across the ocean will send us timely reinforcement to bridge the gap between the peace flotillas of 1939 and war flotillas of 1941. There is no difficulty in sending such aid. The seas and oceans are open. The U-boats are contained. The magnetic mine is, up to the present time, effectively mastered. The merchant tonnage under the British flag, after a year of unlimited U-boat war, after eight months of intensive mining attack, is larger than when we began. We have, in addition, under our control at least 4,000,000 tons of shipping from the captive countries which has taken refuge here or in the harbours of the Empire. Our stocks of food of all kinds are far more abundant than in the days of peace and a large and growing programme of food production in on foot.

Why do I say all this? Not assuredly to boast; not assuredly to give the slightest countenance to complacency. The dangers we face are still enormous, but so are our advantages and resources. I recount them because the people have a right to know that there are solid grounds for the confidence which we feel, and that we have good reason to believe ourselves capable, as I said in a very dark hour two months ago, of continuing the war, 'if necessary alone, if necessary for years.' I say it also because the fact that the British Empire stands invincible, and that Nazidom is still being resisted, will kindle again the spark of hope in the breasts of hundreds of millions of down-trodden or despairing men and women throughout Europe, and far beyond its bounds, and that from these sparks there will presently come a cleansing and devouring flame.

"The great air battle which has been in progress over this Island for the last few weeks has recently attained a high intensity. It is too soon to attempt to assign limits either to its scale or to its duration. We must certainly expect that greater efforts will be made by the enemy than any he has so far put forth. Hostile airfields are still being developed in France and the Low Countries, and the movement of squadrons and material for attacking us is still proceeding. It is quite plain that Herr Hitler could not admit defeat in his air attack on Great Britain without sustaining most serious injury. If, after all his boasting and blood-curdling threats and lurid accounts trumpeted round the world of the damage he has inflicted, of the vast numbers of our Air Force he has shot down, so he says, with so little loss to himself; if after tales of the panic-stricken British crouched in their holes cursing the plutocratic Parliament which has led them to such a plight; if after all this his whole air onslaught were forced after a while tamely to peter out, the Führer's reputation for veracity of statement might be seriously impugned. We may be sure, therefore, that he will continue as long as he has the strength to do so, and as long as any preoccupation he may have in respect of the Russian Air Force allow him to do so.

"On the other hand, the conditions and course of the fighting have so far been favourable to us. I told the House two months ago that whereas in France our fighter aircraft were wont to inflict a loss of two or three to one upon the Germans, and in the fighting at Dunkirk, which was a kind of no man's land, a loss of about three or four to one, we expected that in an attack on this Island we should achieve a larger ratio. This has certainly come true. It must also be remembered that all the enemy machines and pilots which are shot down over our Island, or over the seas which surround it, are either destroyed or captured; whereas a considerable proportion of our machines, and also of our pilots, are saved, and soon again in many cases come into action.

"A vast and admirable system of salvage, directed by the Ministry of Aircraft Production, ensures the speediest return to the fighting line of damaged machines, and the most provident and speedy use of all the spare parts and material. At the same time the splendid, nay, astounding increase in the output and repair of British aircraft and engines which Lord Beaverbrook has achieved by a genius of organization and drive, which looks like magic, has given us overflowing reserves of every type of aircraft, and an ever mounting stream of production both quantity and quality. The enemy is, of course, far more numerous than we are. But our new production already, as I am advised, largely exceeds his, and the American production is only just beginning to flow in. It is a fact, as I see from my daily returns, that our bomber and fighter strengths now, after all this fighting, are larger than they have ever been. We hope, we believe that we shall be able to continue the air struggle indefinitely and as long as the enemy pleases, and the longer it continues the more rapid will be our approach, first towards that parity, and then into that superiority in the air, upon which in a large measure the decision of the war depends.

The gratitude of every home in our Island, in our Empire, and indeed throughout the world, except in the abodes of the guilty, goes out to the British airmen who, undaunted by odds, unwearied in their constant challenge and mortal danger, are turning the tide of world war by their prowess and by their devotion. Never in the field of human conflict was so much owed by so many to so few. All hearts go out to the fighter pilots, whose brilliant actions we see with our own eyes day after day, but we must never forget that all the time, night after night, month after month, our bomber squadrons travel far into Germany, find their targets in the darkness by the highest navigational skill, aim their attacks, often under the heaviest fire, often with serious loss, with deliberate, careful discrimination and inflict shattering blows upon the whole of the technical and war-making structure of the Nazi power. On no part of the Royal Air Force does the weight of the war fall more heavily than on the daylight bombers who will play an invaluable part in the case of invasion and whose unflinching zeal it has been necessary in the meanwhile on numerous occasions to restrain.

"We are able to verify the results of bombing military targets in Germany, not only by reports which reach us through many sources, but also, of course, by photography. I have no hesitation in saying that this process of bombing the military industries and communications of Germany and the air bases and storage depots from which we are attacked, which process will continue upon an ever increasing scale until the end of the war, and may in another year attain dimensions hitherto undreamed of, affords one at least of the most certain, if not the shortest of all roads to victory. Even if the Nazi legions stood triumphant on the Black Sea, or indeed upon the Caspian, even if Hitler was at the gates of India, it would profit him nothing if at the same time the entire economic and scientific apparatus of German war power lay shattered and pulverised at home.

"The fact that the invasion of this Island upon a large scale has become a far more difficult operation with every week that has passed since we saved our Army at Dunkirk, and our very great preponderance of sea power, enable us to turn our eyes and to turn our strength increasingly towards the Mediterranean and against that other enemy who, without the slightest provocation, coldly and deliberately, for greed and gain, stabbed France in the back in the moment of her agony, and is now marching against us in Africa. The defection of France has, of course, been deeply damaging to our position in what is called, somewhat oddly, the Middle East. In the defence of Somaliland, for instance, we had counted upon strong French forces attacking the Italians from Jibuti. We had counted also upon the French naval and air bases in the Mediterranean, and particularly upon the North African shore. We had counted upon the French Fleet. Even though metropolitan France was temporarily overrun, there was no reason why the French Navy, substantial parts of the French Army, the French Air Force and the French Empire overseas should not have continued the struggle at our side.

"Shielded by overwhelming sea-power, possessed of invaluable strategic bases and of ample funds, France might have remained one of the great combatants in the struggle. By so doing, France would have preserved the continuity of her life, and the French Empire might have advanced with the British Empire to the rescue of the independence and integrity of the French Motherland. In our own case, if we had been put in the terrible position of France, a contingency now happily impossible, although, of course, it would have been the duty of all war leaders to fight on here to the end, it would also have been their duty, as I indicated in my speech on 4th June, to provide as far as possible for the Naval security of Canada and our Dominions and to make sure they had the means to carry on the struggle from beyond the oceans. Most of the other countries that have been overrun by Germany for the time being have persevered valiantly and faithfully. The Czechs, the Poles, the Norwegians, the Dutch, the Belgians are still in the field, sword in hand, recognised by Great Britain and the United States as the solid representative authorities and lawful Governments of their respective States.

"That France alone should lie prostrate at this moment, is the crime, not of a great and noble nation, but of what are called 'the men of Vichy.' We have profound sympathy with the French people. Our old comradeship with France is not dead. In General de Gaulle and his gallant band, that comradeship takes an effective form. These free Frenchmen have been condemned to death by Vichy, but the day will come, as surely as the sun will rise tomorrow, when their names will be held in honour, and their names will be graven in stone in the streets and villages of a France restored in a liberated Europe to its full freedom and its ancient fame. But this conviction which I feel of the future cannot affect the immediate problems which confront us in the Mediterranean and in Africa. It had been decided some time before the beginning of the war not to defend the Protectorate of Somaliland, and when our small forces there, a few battalions, a few guns were attacked by all the Italian troops, nearly two divisions, which had formerly faced the French at Jibuti, it was right to withdraw our detachments, virtually intact, for action elsewhere. Far larger operations no doubt impend in the Middle East theatre, and I shall certainly not attempt to discuss or prophesy about their probable course. We have large armies and many means of reinforcing them. We have the complete sea command of the Eastern Mediterranean. We intend to do our best to give a good account of ourselves, and to discharge faithfully and resolutely all our obligations and duties in that quarter of the world. More than that I do not think the House would wish me to say at the present time.

"A good many people have written to me to ask me to make on this occasion a fuller statement of our war aims, and of the kind of peace we wish to make after the war, than is contained in the very considerable declaration which was made early in the Autumn. Since then we have made common cause with Norway, Holland and Belgium. We have recognised the Czech Government of Dr. Benes, and we have told General de Gaulle that our success will carry with it the restoration of France. I do not think it would be wise at this moment, while the battle rages and the war is still perhaps only in its early stage, to embark upon elaborate speculations about the future shape which should be given to Europe or the new securities which must be arranged to spare mankind the miseries of a third World War. The ground is not new, it has been frequently traversed and explored, and many ideas are held about it in common by all good men, and all free men. But before we can undertake the task of rebuilding we have not only to be convinced ourselves, but we have to convince all other countries that the Nazi tyranny is going to be finally broken. The right to guide the course of world history is the noblest prize of victory. We are still toiling up the hill, we have not yet reached the crest-line of it, we cannot survey the landscape or even imagine what its condition will be when that longed-for morning comes. The task which lies before us immediately is at once more practical, more simple and more stern. I hope—indeed I pray—that we shall not be found unworthy of our victory if after toil and tribulation it is granted to us. For the rest, we have to gain the victory. That is our task.

"There is, however, one direction in which we can see a little more clearly ahead. We have to think not only for ourselves but for the lasting security of the cause and principles for which we are fighting and of the long future of the British Commonwealth of Nations. Some months ago we came to the conclusion that the interests of the United States and of the British Empire both required that the United States should have facilities for the naval and air defence of the Western hemisphere against the attack of a Nazi power which might have acquirted temporary but lengthy control of a large part of Western Europe and its formidable resources. We had therefore decided spontaneously, and without being asked or offered any inducement, to inform the Government of the United States that we would be glad to place such defence facilities at their disposal by leasing suitable sites in our Transatlantic possessions for their greater security against the unmeasured dangers of the future. The principle of association of interests for common purposes between Great Britain and the United States had developed even before the war. Various agreements had been reached about certain small islands in the Pacific Ocean which had become important as air fuelling points. In all this line of thought we found ourselves in very close harmony with the Government of Canada.

"Presently we learned that anxiety was also felt in the United States about the air and naval defence of their Atlantic seaboard, and President Roosevelt has recently made it clear that he would like to discuss with us, and with the Dominion of Canada and with Newfoundland, the development of American naval and air facilities in Newfoundland and in the West Indies. There is, of course, no question of any transference of sovereignty—that has never been suggested—or of any action being taken, without the consent or against the wishes of the various Colonies concerned, but for our part, His Majesty's Government are entirely willing to accord defence facilities to the United States on a 99 years leasehold basis, and we feel sure that our interests no less than theirs, and the interests of the Colonies themselves and of Canada and Newfoundland will be served thereby. These are important steps. Undoubtedly this process means that these two great organisations of the English-speaking democracies, the British Empire and the United States, will have to be somewhat mixed up together in some of their affairs for mutual and general advantage. For my own part, looking out upon the future, I do not view the process with any misgivings. I could not stop it if I wished; no one can stop it. Like the Mississippi, it just keeps rolling along. Let it roll. Let it roll on full flood, inexorable irresistable, benignant, to broader lands and better days."

Fifty-Seven Nights

Having failed on a grand scale to secure air supremacy over England in the Battle of Britain, to enable the German invasion of Britain, Hermann Goering's air force was left with the much longer-ranging task of crushing the morale of the British people and coercing them into capitulation through a campaign of nocturnal bombing. The heavy raids beginning in early September were planned, on an alternating basis, to concentrate on the destruction of civilian morale and also the British supply and production centres. This campaign to punish the British was wrought over a period of fifty-seven nights beginning on 7 September and continuing through the following May, and has become known as the Blitz. In it an estimated 41,000 civilians were killed and nearly 140,000 injured, with about half those numbers in London. More than 1,000,000 homes were destroyed or heavily damaged and enormous damage was done to military-related and other British industry. In their targetting, the Germans were attracted to the Birmingham area where local factories were heavily engaged in tank and Spitfire production, and to the munitions production in Coventry. But British war production was never seriously reduced in the eight-month bombing campaign, nor were the British people significantly demoralized or moved to want to surrender. Importantly, war production was significantly expanded through and after the Blitz. The Blitz campaign of bombing in Britain, while punishing and extremely serious in the casualties and damage it caused, did not compare with the loss of life and the scale of devastation brought by the Allied combined bombing offensive against Germany, in which the bombing of Hamburg alone created more than 42,000 casualties.

By November the bombing campaign had been extended to other British cities and, for the most part, the enemy raiders came and went on their deadly night-bombing assignments with very little response or interference by local or night-fighter defence. The new British Airborne Interception radar system was about to become operational against the night raiders, but had not yet been put to the test and was still essentially experimental. From the German perspective, the campaign of night-bombing was under way at a lower rate of efficiency than was customary for the Luftwaffe. Its night navigation and bombing accuracy were not up to standard, which was largely due to it having lost so many of its more experienced bomber crews in the run-up to and early part of the Blitz.

The Luftwaffe bomber crews at that point were dependent upon three main blind-bombing and navigation systems. The first was called *Knickebein* (bent leg) and was utilized in bomber aircraft equipped with a Lorenz blind-approach device. With this system, a pilot flew his bomber along a radio beam to the target. His bomb aimer released the bombload when his aircraft reached a second radio signal that crossed the approach beam. Advantage: the system could be employed by a large number of aircraft. Disadvantage: the system was easily jammed. The second such system was called *X-Gerät* (X Equipment) and was made specifically for the Heinkel He 111H-4 bombers of Kampfgruppe 100. This system comprised one approach beam that was crossed in the vicinity of the target by three other beams. The pilot received the first such signal some thirty-one miles short of the target, at which point he accurately aligned his aircraft on the approach beam. The second and third traversing beams were reached at twelve miles and three miles from the target respectively. A system computer then used the groundspeed of the aircraft to set an automatic bomb release point when the third beam signal was received. Advantage: highly accurate. Disadvantage: the system could be easily jammed.

Searchlights illuminating the cathedral of St Paul's in the city of London during the Blitz of 1940-41.

English mothers and grandmothers are shown the use of a gas mask contraption designed to protect babies.

The third such system used by the German bomber crews early in the Blitz was *Y-Gerät* (Y equipment), a bombing aid used only in the He 111s of III/KG 26. In this system, bomb release occurred at an accurately-measured point along an approach beam. The range computation was made by a ground radar station which sent a pulse signal to the bomber. After an established interval the bomber returned a pulse to the ground station and the range to the bombing point was then computer-calculated in milli-seconds. The ground station then relayed the bomb-release signal. Advantage: very accurate. Disadvantage: readily jammed.

Night navigation and bombing aid accuracy were not the only problems that the Luftflotten 2, 3 and 5 had to cope with in their Blitz attacks on London and the other major cities of Britain. In theory, the Luftwaffe had upwards of 1,300 bombers available for use in the attacks, but in practice only about 700 were actually serviceable and employed in the raids. In the month that followed the huge daylight raid of 7 September on the London docks, an average of between sixty and 260 bombers were utilized each night. The Knickebein blind-bombing and navigation system was the primary system in use during October, and was effectively jammed by RAF Signals personnel, reducing the navigation and bombing capability of the Germans to reliance on moonlight. From 9 October, the German bomber fleets were ordered to raise the bomb tonnage they were delivering, in an increased effort to panic the British civilian population and force a capitulation. But, just as they had failed in securing air supremacy over England in the Battle of Britain, they failed too in this effort.

That failure led to an expansion by Goering of the present bombing campaign to one in which a lengthy attrition of British war-related industry was the goal. It was then that the X-Gerät navigation and bombing system entered service with the bombers of KGr 100. It was utilized in conjunction with the dropping of incendiary bombs for target-marking purposes in aid of the main bomber force to follow in to the target area. In an early example of the X-Gerät system deployment, KGr 100 attacked the city of Coventry on the night of 14 / 15 November. In the attack, a dozen Heinkel He 111s dropped more than 1,000 incendiaries on the city at 8:15 p.m. to mark it for the three separate main bomber streams then approaching from the directions of the Wash, Portland, and Dungeness. The terrifying raid lasted through the night, with wave after wave of bombers—469 sorties in all—blasting Coventry with nearly 400 tons of high explosives, as well as fifty-six tons of incendiaries and 127 parachute sea-mines. Once again, there was little in the way of air defences; an easy night for the raiders. Not so easy, however, on the night of 19 November, when the German bomber force attacked Birmingham in a heavy raid that cost the attackers five of their aircraft. Keeping up the large-scale attacks through the rest of November and December, the Luftwaffe struck heavily at Southampton, Sheffield, Liverpool, Bristol, Plymouth, and London. Only the relatively sudden appearance of rapidly deteriorating weather put a stop to a massive raid on London in the night of 29 December 1940 after two hours of attacking. The 130 sorties flown that evening were all incendiary deliveries and, despite the early curtailment of the raid, by 10 p.m. the number and scale of the raging fires in the City area of London exceeded the scale of the Great Fire of 1666. The bulk of the incendiaries fell just to the northwest of St Paul's Cathedral and the fires they started served to guide the delivery of the high explosives of the main bomber force that followed.

By January 1941, the Luftwaffe had developed a distinct lack of confidence in the X-Gerät system of blind-bombing aid, and that, in combination with the RAF's ability to jam the system, together with its growing use of decoy fires, helped to force the Germans to navigate and bomb mainly by moonlight and to again redirect their targetting now to the British seaports of Swansea, Cardiff, Bristol, Hull and Plymouth. As the winter progressed, the Luftwaffe was having to redeploy much of its bomber force to the Mediterranean, and to take measures aimed at making the British believe that more German bombers were taking part in the raids on Britain than was actually the case. In order to project the impression of larger-scale bombing attacks into the spring, the Germans mounted the largest raid of the Blitz on the night of 10 May, with many of the bomber crews flying up to three sorties that night in the delivery of nearly 800 tons of high explosives and some 87,000 incendiaries. Immense damage was inflicted on greater London in this raid, and another of at least equal size and intensity was levelled on the British capital two nights later.

As May dragged on, it became obvious that the German air force was radically redirecting its forces from the west, to bases in Eastern Poland and Germany in preparation for the coming major

St Paul's cathedral by daylight during the Blitz.

offensive against the Soviet Union. Clearly, the Blitz on Britain was drawing down and those German air units remaining in France were mostly limited in their activity to anti-shipping attacks and mine-laying. Again, the Luftwaffe had wasted the opportunity it had been given to force surrender from Britain through adherence to failed bombing policy, and having allowed the most heavily-bomb-damaged major cities and industrial areas of Britain to recover from their attacks. The final position taken by the Germans towards Britain was one of doing all they could to starve the British into submission through the continued, long-term destruction of the food and supplies destined to her by sea and air shipping, a position the Germans now anticipated to follow their expected victory over the Soviets in late 1941. Hitler had believed that his air force could terrorize the people of London into defeat and surrender. Instead, the Londoners displayed a rather perverse pleasure in taking their place in the front lines of the war. Their motto became: We can take it!

Any consideration of the German air offensive against Britain must go into the history of the Luftwaffe and strategic bombing. Since the 1920s, when the best-known air power theorists—General Billy Mitchell, General Giulio Douhet, and Generalleutnant Walther Wever—were projecting their views on the ability of modern air forces to win wars without the traditional requirement of land and sea forces, many in military circles the world over accepted the notion that there was no effective defence against air attack. There was also wide-spread acceptance in military and political circles of the theory that the heavy bombing of civilian residential areas would lead to the collapse of the people's will to continue the war and war production. Indeed, the bombing policy of RAF Bomber Command through much of the Second World War was based largely on what the Germans themselves later branded as 'terror bombing', the targetting of civilian population areas along with industrial and communications objectives. Luftwaffe planners, while accepting the concept of strategic bombing of enemy industry and cities, disagreed with the theory that the use of air power alone could be war-winning. They failed to develop a sound, long-term strategy for an efficient bombing campaign to destroy British war industry and the Royal Air Force. When Walther Wever was the Luftwaffe Chief of the General Staff in 1935, he stressed the significance of the strategic bomber and developed an outline for a new air strategy:

1. To destroy the enemy air force by bombing its bases and aircraft factories, and defeating enemy air forces attacking German targets.
2. To prevent the movement of large enemy ground forces to the decisive areas by destroying railways and roads, particularly bridges and tunnels, which are indispensable for the movement and supply of forces.
3. To support the operations of the army formations, independent of railways, i.e. armoured forces and motorized forces, by impeding the enemy advance and participating directly in ground operations.
4. To support naval operations by attacking naval bases, protecting Germany's naval bases and participating directly in naval battles.
5. To paralyse the enemy armed forces by stopping production in the armaments factories.
In Wever's view, the General Staff of the Luftwaffe needed to be more well-grounded in the areas of war economics, grand strategy, armaments production, and the art and science of understanding the enemy, all this in addition to tactical and operational concerns. His view was never adopted. With his death in an air crash in 1936, his successors, Albert Kesselring and Hans-Jürgen Stumpff, both advocates of a narrower tactical structuring of the Luftwaffe, chose to shape it essentially on the air sup-

port role. They were themselves supported in that perspective by Hugo Sperrle and Hans Jeschonnek, veteran German airmen.

Oddly, the offence-minded Adolf Hitler seemed less interested in the planning and details of bombing Britain into submission than he was in the protection of Germany's cities and war production facilities from enemy bombs. In the late 1930s he had advocated strategic bombing and had talked with Luftwaffe planners about eventually using the German bomber force to crush the will of the British people. Later, during the Blitz attacks, he became quite negative about the bombing results achieved by his air force: "The munitions industry cannot be interfered with effectively by air raids. Usually the prescribed targets are not hit." He appeared somewhat disinterested in the development of the strategic bombing campaign against Britain and declined to take a firm hand in it. He seemed to be more enthusiastic about the use of bombing as a terror weapon to wreck the civilian morale than as a realistic means to effective economic warfare and the elimination of the enemy's war-making capacity. He seemed to believe that his 1930s-era adherence to morale-breaking through bombing would sufficiently intimidate the enemy from entering into a policy of unrestricted bombing against German targets. He appeared to be more politically concerned with protecting the German population from Allied bombing, to ward off the the threat of popular revolt against his regime. That concern played a part in his ramping up the terror-bombing against the British in the anticipation of creating some sort of stalemate wherein both the British and Germans would pull back from the bombing policy.

Hitler also was contending with the arrogant and egotistical Goering whose proprietary approach to his Luftwaffe force caused him to resist every attempt by Admiral Erich Raeder to operate and control a "naval air force" of his own for the German Navy. Hitler fended off those requests with: "We should never have been able to hold our own in this war if we had not had an undivided Luftwaffe." Goering operated the Luftwaffe as his own little empire, a free rein allowed him for the most part by Hitler, who in later stages of the war would find it more and more difficult to intervene in the direction of the air force. This was exacerbated by Goering's habit of misrepresenting and over-optimistically interpreting operational results and strength capabilities to the German leader. In trying to manage the news about the Luftwaffe for Hitler's consumption, he went as far as to stage an air display in the summer of 1939 in an effort to show the Führer that his air force was better prepared for strategic air war than it actually was.

As planning for the Blitz on Britain evolved in late August 1940, the Luftwaffe General Staff concentrated initially on daylight attacks against the enemy's key industrial cities, beginning with the London attack of 7 September. The results of the major raid of 15 September, in which large air battles occurred, were quite punishing for the German air force and produced little real gain for them. A few more significant daylight raids followed between mid-September and into October, but by then the Luftwaffe had decided to minimize its losses and switch to night attacks, official Luftwaffe policy from 7 October. One factor in its inability to cause much greater damage than it was achieving to that point was, of course, the limited bombload capacity of its principle bomber aircraft, the Heinkel He 111, Junkers Ju 88, and the Dornier Do 17. In the years between the Great War and the Second World War, the Luftwaffe elected to pursue medium bombers for its inventory for three fundamental reasons: 1) The Luftwaffe General Staff was persuaded that a medium-bomber force was as capable of carrying out strategic bombing missions as a heavy-bomber force; 2) The Germans did not then have either the technical ability or the resources to build an effective four-engined strategic bomber; and 3) Hitler did not plan or project a war with Britain in 1939.

In addition, the General Staff had not been told to even consider Britain as a potential war opponent until 1938; had not really begun to research and gather sufficient intelligence about British industry; and seemed utterly incapable of agreeing an appropriate strategy of attack. The logic and reason of their eventual bombing strategy grew ever more confused and apparently aimless through the winter months of 1940-41. The British were, on the whole, ahead of the German enemy in anticipating, understanding and countering aspects of the bombing offensive, particularly in terms of dispersing military-related production facilities to reduce their exposure and vulnerability to any concentrated bombing attack. A very effective network of cottage industry was developed at that critical time and an efficient capability for the distribution of supplies, spares, parts and equipment was organized to support the growing British war economy.

By late 1938 and early 1939, many people in the British government and elsewhere in the country had come to believe that war with Germany was virtually inevitable. Experts and so-called authorities in various fields were quick to make predictions about the impact of such a war on the British people. Winston Churchill, in a 1934 speech to Parliament had said: "We must expect that, under the pressure of continuous attack upon London, at least three or four million people would be driven out into the open country around the metropolis." War-related rumours were rampant and in 1939 the government developed a plan for the voluntary evacuation of four million people, mainly women and children, from London and other major urban areas. The plan called for ninety percent of the evacuees to be housed in private homes, and a survey was conducted to define the available households for use in the implimentation of the plan.

On 10 August 1939, a "black-out" exercise was held and, from 1 September, the day that Germany invaded Poland, Britain was plunged into a compulsory six-year-long black-out at sunset, a condition viewed by most Britons as even worse than the deprivation of rationing, which actually lasted into the 1950s. Another consideration of the period was the availability and preparation of air raid shelters nation-wide for the populous. Such arrangements became the responsibility of many local government authorities, especially in the most threatened communities like Birmingham, the East End of London, Coventry, Plymouth, Portsmouth, Southampton and Belfast, all of which lacked sufficient adequate shelter facilities. Fortunately, the combination of the "Phoney War" period and the delayed beginning of the German bombing campaign on Britain's cities and industry, enabled some of the targetted communities to build and arrange for such sheltering facilities.

But the problem of sheltering the people of Britain from the bombs of the Luftwaffe was a big one. Many people in a target city like London naturally looked to the city's Underground subway system with its deep stations for such protection. The British government and the London authorities were initially opposed to the use of those deep subway stations as bomb shelters. Though many Londoners had used them as shelters during the First World War, the authorities early in WW2 refused to let them be used as such on the grounds that it would interfere with commuter and troop travel, and that the shelterers might simply refuse to leave. After the second week of September 1940, however, the government relented and the people of London were allowed to shelter in the Underground stations. The greatest number of people to use the deep shelters was 177,000 on the night of a heavy raid, 27 September. Surveys showed that an average of about four percent of London residents used the deep shelters during the heaviest raids of the Blitz, while some twenty-seven percent utilized Anderson backyard shelters and, later, Morrison in-door shelters in private homes. Both types of private home shelters were distributed by the government, but many of the Anderson shelters were later abandoned as unsafe. By the time the government yielded to public demand for the building of new deep shelters in the Underground system, the shelters were not completed and ready for use until the heaviest bombing had ended. By then, the government had provided bunks in the larger shelter areas, as well as stoves, bathrooms, and canteen trains to offer food. Timed tickets were issued for the bunks in the larger shelters, which reduced the amount of time spent queuing for them. Both the Salvation Army and the British Red Cross were involved in efforts to improve the conditions in the shelters. Entertainment was even provided in some of them. It included films, concerts, plays, and books from the libraries.

The emotional and mental pressures endured by Britons in the Blitz caused many problems ranging from extraordinary fatigue, anxiety, eating disorders, a variety of mental conditions, miscarriages, and other health problems. But a special network of psychiatric clinics that opened in the period to deal with mental concerns was ultimately closed due to under-use. In general, British morale was high and remained so throughout the Blitz period, this in spite of the negative war news and the threat of invasion. A mere three percent of Britons thought that their country would lose the war and, in October 1940, eighty-nine percent of the British public approved of the Churchill government. British workers worked longer shifts and worked weekends as well. Contributions to the £5000 Spitfire Funds rose dramatically. Volunteerism grew impressively through the Blitz, with thousands joining the Air Raid Precautions Service, the Auxiliary Fire Service, the Home Guard and the Women's Voluntary Services.

In the area of air defence, the primary emphasis had long been on daylight defences, the province of Fighter Command, with little effort and resources expended on night air defence. The RAF was still dominated by those who adhered to the doctrine of the early Chief of the Air Staff, Lord Trenchard, who believed that offence was the best defence, "the cult of the offensive." The basis of this idea held that the prevention of German bombers hitting targets in Britain depended upon RAF Bomber Command hitting and destroying the enemy air force on its own bases, and the enemy aircraft factories, as well as its fuel reserves stored at the German oil facilities. The flaw in this idea lay in the fact that RAF Bomber Command was not yet equipped with either the aircraft or the navigation and bombing technology to carry out such assignments, nor would it be for quite some time. Thus, the air defence of Britain, by day and night, was left largely to Fighter Command, which could not do the entire job on its own. An effective night-fighter force did not yet exist at the start of the war and the ground anti-aircraft batteries were not all that effective. By late November, Air Chief Marshal Dowding, Air Officer Commanding Fighter Command, had been replaced by William Sholto Douglas who, in 1946, would be promoted to Marshal of the Royal Air Force.

It was only after the main action of the Blitz had ended, in May 1941, that a combination of relatively effective anti-aircraft defences and a useful night-fighter force was able to make a difference. Until then, the main force of German bombers had little difficulty in getting through local defences to their targets. The British had to resort to a variety of ruses in the interim in an attempt to divert the enemy aircraft from their targets; these included dummy airfields, simulated industrial and residential areas with artificial street lighting, decoy fires and a range of other diversionary tricks.

As the Blitz continued through the autumn, however, German operational losses mounted and, by 7 October, Hermann Goering had decided that they were unsustainable relative to the results being achieved. He then switched the Luftwaffe to night bombing attacks on Britain. By the end of October, more than 13,000 civilians had been killed in London, with nearly 20,000 injured.

After November 1940, the Luftwaffe broadened the scope of its night bombing of Britain, going after additional targets especially in the West Midlands with particular emphasis on Birmingham and Coventry. But throughout the fifty-seven most brutal nights of the Blitz, there was no real let-up on London, and by May 1941 the civilian casualties in the capital amounted to 28,556 killed and 25,578 injured. As the first few months of 1941 passed, the toll on the German bomber force became obvious. The serviceability of its bomber aircraft fell to the point where just 550 of 1,215 such aircraft were available for operational use.

In a final pair of major attacks of the Blitz on London during the nights of 10 / 11 and 11 / 12 May, the Luftwaffe bombers dropped 800 tons of explosives and incendiaries, causing more than 2,000 significant fires. In the second raid, 1,436 people were killed, with 1,792 injured. In the attack, the Law Courts and Westminster Abbey were severely damaged and one chamber of the House of Commons was destroyed. By late May, the sporadic bombing policies of the Luftwaffe General Staff had clearly doomed the bombing offensive against Britain. It was clear too, that by May the attentions of the Germans had shifted heavily towards the east and the Soviet Union. They had achieved relatively little strategically against the British apart from causing great infrastructure damage, killing about 41,000 Britons, and injuring upwards of 139,000. The British felt that they had learned some valuable lessons from the experience of the Luftwaffe in the Blitz. They noted that incendiaries appeared to have a greater effect than high explosives on enemy industrial targets; they were impressed by the devastation wrought on city centres and the effects on transport, utilities, and administration; they concurred on the staggering effect of the incendiaries in area attacks on enemy cities; on the validity of striking a single target each night with maximum bomb tonnage for maximum impact; and perhaps most importantly, they recognized that the morale of the British people had not been broken by the intensive bombing campaign.

A lobby card poster advertising the film *The First of the Few*, a 1942 release that starred Leslie Howard and David Niven. It told the story of Spitfire designer R.J. Mitchell and how the great Second World War fighter came about.

Payback Time

The British government first began getting reports of unusual enemy activity on the Baltic coast in the spring of 1943. Their photo reconnaissance of the peninsula was showing newly-constructed laboratories, a living site and examples of what photo interpreters considered to be missile launching pads. It was known then that German atom bomb research had made relatively little progress to that date and that they were concentrating heavily on the development of both pilotless bombers and long-range rockets at Peenemünde, "the summit of research and development," as Churchill referred to the complex. Throughout the early summer, British intelligence material about the "V-weapons" continued to accumulate, indicating that Hitler intended to begin the new bombardment of London on 30 October. He expected that, by the end of the year, London would be devastated, that the British would have surrendered, and the bulk of German military might could be focused on the Soviet Union. This new German threat against London was considered real and serious, and caused the British government to look again at the evacuation plans it had made in 1939. At the same time, the government decided that a maximum effort should be mounted to eliminate Peenemünde as a specific threat. That project, code-named Hydra, was handed to Air Chief Marshal Arthur Harris, head of RAF Bomber Command.

Harris ordered, from his bunker HQ at High Wycombe, that the attack would be mounted at full strength and in full moonlight, despite the advantage to the German fighters. He said that his bomber crews were not to know the nature of the target—only that it had to be eliminated no matter how difficult the job. The bombing altitude would be between 8,000 and 12,000 feet—less than half the normal height, and a "master of ceremonies" would control the attack from over the target for the first time in a full-scale RAF raid, and that new, slow-burning Newhaven "spot fires" would be used as target markers. All across eastern England, fully 600 bomber crews were assembled in briefing rooms on the afternoon of Tuesday, 17 August.

Sergeant Jack Currie: "One of those crews was mine. We had come together in the usual random manner, responding to a call of 'Sort yourselves out, chaps', in an echoing hangar at the operational training unit. Within five minutes, a bomber crew was formed: three bright Australians as navigator, bomb aimer, and rear gunner, a quiet Northumbrian as wireless operator, and me. Later, converting to the Lancaster, we added two teenagers: a Welshman as mid-upper gunner and a Merseysider as flight engineer. At least I was no longer the youngest in the crew.

"We were assigned to Number Twelve Squadron, at Wickenby, near Lincoln, and they seemed to need us: they had lost four crews in seven days. Ahead lay a tour of thirty operations, and the chances of survival were roughly one in four; they improved, said the old hands, if you got through five missions. We did that, and another three; now we were ready for our ninth.

"I had come to trust the aeroplane and to know the crew. Jim Cassidy, having quietly used a sick-bag as soon as we were airborne, would navigate us to Germany and back with no further trace of frailty. He had always set his heart on being a navigator, unlike many who first aspired to be pilots; he had come out top in training and it showed. Larry Myring, with whom bloody was an all-purpose, mandatory adjective, would complain about the cold and be happy only when the target came in sight. The gunners Charlie Lanham and George Protheroe were always constantly alert; up to now, they had not been required to fire their guns in anger. Charles Fairbairn would be heard only when some-

above: A still from the film *The Dam Busters*, with Richard Todd, right. and Robert Shaw; far left: Air Chief Marshal Arthur Harris, headed RAF Bomber Command in the Second World War; left: A bomb aimer in a Lancaster heavy bomber.

above: 4,000-pound 'cookies' ready for loading aboard Lancaster bombers particpating in Bomber Command's massive offensive against Germany; left: Lancaster forward fuselage sections in an Avro production facility in 1942; top right: Fusing bombs on a hardstand of a U.S. Army Air Force base in England.

thing urgent—a recall, a diversion, or a change of wind—came through on the radio, and Johnny Walker would do what was needed to conserve the fuel. My responsibility as captain, was to make the big decisions—like which dance hall or cinema we went to on a stand-down night.

"The Hydra briefing started with a little white lie. The enemy, said the intelligence officer, were developing a new generation of radar-controlled night fighters on the Baltic coast. That was the carrot. The squadron commander took over with the stick. If we failed to clobber Pennemünde tonight, we would go again the next night and the next until we did. The attack, he continued, would comprise three ten-minute waves: the first would hit the scientists' living quarters, the second wave the airfield (in reality, the rocket-launching sites), and the third the laboratories.

" 'Hey, skip,' Myring whispered, 'What's our squadron motto?'

"I glanced at him. 'You know perfectly well.'

" 'Yeah. *Leads the field.* So how is it we're always in the last bloody wave?'

"The PFF would employ the Newhaven method, which meant visual ground marking—and would 're-centre' the markers on each successive aiming point. We were to listen out on channel C for the

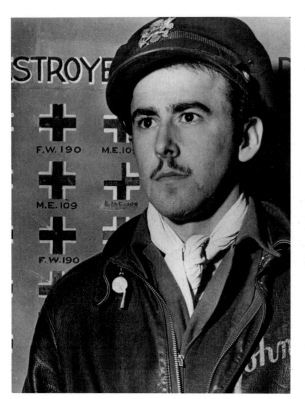

top left: Captain John Godfrey; top centre: Major Howard 'Deacon' Hively, both members of the 4th Fighter Group at Debden, England; above: Major John England on his return from a combat mission in *'Nooky Booky IV'*, in his 357th Fighter Group P-51 Mustang; left: the entire crew of a B-17F bomber arriving at the aircraft for a mission to a German target.

MC's instructions, and follow them to the letter. Purely as a precaution, in case the markers should be temporarily obscured, we were to approach the target on 'time-and-distance' runs from Cape Arkona on Rügen Island, forty miles north of Peenemünde. The outbound route would keep us clear of known flak concentrations, and the target defences were expected to be light. And, as for the enemy night fighters, they would be diverted by no less than eight Mosquitoes bothering Berlin at the time of the attack.

"After the navigation leader had specified the courses, heights, and airspeeds, the weatherman performed his magic-lantern show of cloud tops and bases. 'Looks good,' said Cassidy. 'Larry should get plenty of visual pinpoints.' He looked meaningfully at the bomb-aimer, whose map-reading ability he had sometimes questioned. Myring then grunted: for him, the main business of the briefing began only when the bombing leader took the stage. He licked his pencil, and made a careful note of how his five-ton load would be disposed.

"The signals leader spoke in an apologetic undertone: 'I would like to see all wireless operators for just a few minutes after briefing.' I leaned across to ask Fairbairn 'What's all the secrecy, Charles? Why can't he tell everybody?'

" 'It's just technical stuff, Jack. A pilot wouldn't understand.'

"The veteran gunnery leader, with a battered service cap worn at an angle, advised constant vigi-

The primary fighter types used by the American air force in World War Two England—left: The Lockheed P-38 Lightning; below: the North American P-51D Mustang; bottom: The Republic P-47 Thunderbolt.

lance. Defying popular belief, he saw the moonlight as being to our advantage: 'A fighter will stick out like a sore thumb. Just keep your eyes peeled and make sure you see him before he sees you.'

"The station commander strolled onto the stage, one hand in his pocket, the other smoothing a sleek, dark moustache. He was sure he didn't have to emphasize the importance of the target, and anxious that there should be no early returns. 'Your flying meals will be ready at nineteen-thirty hours, transport to the aircraft at twenty-fifteen. Good luck, chaps.'

"The trouble with the Lancaster, apart from being cramped for space and deathly cold at altitude, was that it had a tendency to swing left on takeoff. If you overcorrected, it swung back to the right, and the more you tried to straighten up, the more it deviated. The trick was to eliminate the swing by leading with the throttle of the port outer engine until the speed was high enough to get the rudders in the airflow for directional control. That was what I did at thirty minutes after nine, and PH-George 2 climbed away at maximum boost and 2,850 rpm. At fields all over Lincolnshire, in Yorkshire, and in East Anglia, 595 pilots did the same. Theoretically, if every aircraft stayed within a two-mile radius of base, their climbing orbits should never coincide: in practice, they occasionally did, and we took precautions. Apart from the navigator, busy at the gee-box, every man kept watch.

"We reached 8,000 feet in under twenty minutes, and that was not too bad. Climbing in a circle wasted lift and thrust: aeroplanes climbed better in nice straight lines. When George 2 was straightened out on course, she gained another 3,000 feet in the next five minutes (it was a curious convention with aircraft as with ships that, no matter how obviously masculine their names, they were always female to their crews.)

"Beneath us there was nothing to be seen. The coastal crossing point at Mablethorpe did not exist as houses, streets, and shoreline, but as 53.20N 00.16E on the navigator's chart. At two minutes before ten, I switched off the navigation lights and the IFF, and George 2 headed out across the unseen waters in a straight line for 'point A', seventy miles west of the North Frisian Islands.

"Skipper from mid-upper, okay to test the guns?'

" 'Go ahead.'

"The Lancaster trembled as the gunners fired short bursts from their Brownings, and the sharp smell of cordite filtered through my mask. Climbing steadily, I followed the eternal visual routine: clockwise round the panel, clockwise round the sky. 'Cultivate the roving eye, Coo-ree,' were the words that 1st Lt Sena, USAAC, had constantly intoned when he was teaching me to fly the Vultee BT-13. Then, there were only 'needle, ball, and airspeed' and the clear, blue sky of Georgia for the eye to rove around, but the principle held good.

"I leveled out at 18,000 feet, and Walker brought the pitch back to cruising rpm. George 2 reached point A at twenty minutes after eleven. 'Pilot from navigator, turn onto one-zero-seven magnetic.'

" 'One-zero-seven. Turning on.'

" 'I'm losing the gee signals now, but I think we're pretty well on track.'

"I conjured up a mental map of the Frisian Islands. 'Let's try not to overfly Sylt, Jim.'

" 'No. She'll be all right, Jack. We should be well north of it.'

"Someone switched a mike on. With two of the crew, I knew who it was before he spoke: Myring, because his mask never seemed to keep the slipstream noises out, and Fairbairn, because he always gave two low whistles—pshew, phsew—to check that he had switched from radio to intercom. This was Myring. 'Can't we have the bloody heat up a bit? It's cold enough down here to freeze the . . .' I said, 'No chatter, Larry,' but I had to admit he had a point. The aircraft's fabric wore a ghostly film of frost; ice crystals sparkled on the aerials and guns. I was beating alternate hands on knees to retain some sense of touch, and was starting to lose contact with my feet.

"'Wireless op, is the cabin heat on full?' Cassidy answered for him. 'He's on WT, listening out for the group broadcast.'

"'Aw, it's pretty well up now, Jack.'

"'Not enough. The outside air temperature is less than minus twenty.'

It was all right for Fairbairn. He sat nearest to the hot air vent. He could have flown in his underwear if he didn't have to move around the aircraft now and then. Cassidy was all right too. His navigation desk, forward of the wireless set, was never really cold. Walker and I, farther forward in the cabin, were only ever warm when we were flying at low altitude on a summer's day, and Myring's station in the nose was colder than a tax collector's heart.

"'Pilot from navigator. Enemy coast coming up in two minutes.'

"'Roger. Let's have an intercom check. Bomb-aimer?' 'Loud and clear, skip.'

"'Wireless op?' 'Phsew, phsew. Strength five.'

"'Mid-upper? . . .'

"At eleven-fifteen, the moon showed its head above the northeastern horizon. Myring, mindful of the navigator's hint, came through on the intercom. 'Bomb-aimer, Jim. We're crossing the coast now— fairly well on track.'

"'I need it exactly, Larry.'

"'Oh yeah. Well, about fifteen miles south of Esbjerg, or whatever they call the bloody place.'

"'Lat and long, Larry, and the time.'

"'Christ, Jim. It's not that easy down here, trying to use me bloody torch an' look at the bloody map, and—'

"I cut in. 'Do your best, bomb-aimer.'

"As George 2 headed east-southeast across the southern plains of Denmark, Lanham called from the rear turret: 'Flak astern of us, ten degrees to starboard.'

"'Roger.' Some airmen behind us were a little south of track, and the guns of Sylt were making them aware of it.

"Walker checked the fuel. George 2 was half a ton lighter than her takeoff weight.

"'Let's reduce the revs a bit,' I suggested.

"'We should be able to maintain airspeed at twenty-three hundred.'

"Walker inched the levers down, and I turned the elevator trim minutely back. George 2 reacted badly to the trivial economy. Her indicated airspeed fell by nearly five miles per hour, and she handled like a ship without a rudder.'

"'No go, Johnny, try twenty-three-fifty.'

"Ten minutes passed while we struggled to regain George 2's good will. Five minutes before midnight, Myring came through with a pinpoint; ten minutes later he produced another. 'Fifty-four forty-eight north, twelve thirty-nine east, time . . . er, bloody hell
. . . zero-zero-zero-five.'

"'Good on yer, Larry,' said Cassidy.

"I began a gradual descent to the bombing height as George 2 crossed the southern Swedish islands and continued east-southeast. The moonlight showed the shapes of other bombers, well scattered, moving like a ghostly skein of geese across the Baltic Sea.

left: Captain Richard Bynum, 388th Bomb Group; right: Staff Sergeant S. J. Malinowski, 385th Bomb Group, 8th USAAF

" 'The wind's picked up a bit,' said the navigator. We'll be early at point C. Can you reduce the speed?'

" 'Can't go much slower than one-fifty, Jim. I can make an orbit. How much time do your want to lose?'

" 'Three minutes ought to do it.'

"It was not a manoeuvre I would have cared to make in total darkness or in cloud but, in the moon-light, it was not too hazardous. While the crew watched out, George 2 completed a wide descending circle without bumping into any other aeroplanes. Shortly before we reached the turning point at Cape Arkona, Myring saw the first green spot fires burning on Peenemünde. The time was seventeen minutes after midnight, and we had fifty miles to run.

"The last wind velocity Cassidy had found was from 290 degrees at forty miles per hour. Myring fed the numbers into the bombsight's calculator. Now the curve of the coast was perfectly distinct, and a white line of surf showed on the shore. The scene ahead was much less serene. The detonating 'cookies' made bright, expanding circles, like heavy stones dropped into pools of liquid fire; search-lights were waving, flak bursts were twinkling, and fires were taking hold. Not so many searchlights, nor so much flak as we were accustomed to, but a lot of smoke—more than the fires would seem to justify. The people on Peenemünde were putting up a smoke screen. It was a nice try, and it might have been successful—if the wind hadn't blown it out to sea.

"From 9,000 feet, in the light of the full moon, the target was much closer, and warmer, than the norm. I told Fairbairn to reduce the cabin heat, and Walker turned the oxygen to 'high.' On the starboard beam an interlacing pattern of tracer bullets appeared and disappeared. The voice of the MC sounded on the radio, the cool, clear voice of someone accustomed to command. It was strange, and rather comforting, to hear that English voice in the headphones, high above the Baltic, six hundred miles

Captain Don Gentile, with his crew chief and their P-51 Mustang 'Shangri-La' at Debden, Essex, England.

from home. 'Come in, third wave, and bomb the centre of the green TIs. Let's have a good concentration. Aim right at the centre of the greens.'

" 'Switches on,' said Myring, in the special, growling tone he adopted for the moment—his moment of all moments. 'Bombs fused and selected.'

"I took a deep breath and a fresh grip on the wheel. A spatter of light flak danced around George 2: I tried to pretend it wasn't there. 'Running in nicely, skip,' said Myring, 'Steady as she goes.'

" 'Third wave, don't bomb short. Make sure you aim at the centre of the greens.' I turned down the volume; from now on Myring was in charge.

" 'Bomb doors open, skip.'

"I pushed the lever. The roar of the slipstream made a deeper sound as the bombbay gaped. George 2 tried to raise her nose, and I stilled her with the trim tab. 'Lanc on fire at four o'clock level,' said Lanham. 'It's going down.'

" 'Steady,' Myring growled, 'Left, left . . . steady, steady . . . a touch left, and steady, steady . . . bombs gone!'

"Down went the 4,000-pound 'cookie' six 1,000-pounders, and the two 500-pounders. George 2 lifted as they fell. Cassidy logged the time: thirty-eight minutes after midnight. 'Bomb doors closed,' called Myring. 'Steady for the camera.' In the last few minutes he had uttered thirty words without a single bloody. He really was a changed man with a bomb-tit in his hand.

" 'Bomb doors closed' I echoed, pulling up the lever and rolling back the trim. Thirty seconds passed while the photo plates ticked over, and before I turned George 2 away, away from the brightly burning debris that was Peenemünde. 'New course, Nav?'

" 'Two-nine-five degrees magnetic.'

"Turning on. Give me twenty-eight-fifty, Johnny, and we'll grab some altitude.'

"George 2 climbed away smoothly and headed to the west. We had no way of knowing that the *Nachtjagd* controllers, aware now that the Berlin raid was no more than a feint, had redirected all their available Messerschmitts and Junkers to our homeward route.

"The Lancaster's electronics included a receiver that picked up transmissions from the Lichtenstein radar sets in the German fighters. The radar device was code-named Boozer, perhaps because the red lamp it lighted on the panel was reminiscent of a heavy drinker's nose. At 18,000 feet over Stralsund, thirty miles west of Peenemünde, the roving eye picked the glow up straight away.

" 'Rear gunner from pilot, I have a Boozer warning.'

" 'Rear gunner watching out astern.'

"Boozer also read transmissions from the ground-based Würzburg radars, which could be quite a nuisance when you were flying in the stream; at all times, however, you had to heed the signal. It was as well we did: seconds later, Lanham spoke again. 'Fighter at seven o'clock low. Stand by to corkscrew.'

" 'Standing by.'

" 'Mid-upper from rear gunner. There could be a pair. I'll take care of this one, you watch out.'

" 'I didn't like the sound of that remark. It would be difficult enough to evade one fighter in the moonlight, let alone two. I sat up straight, and gently shook the wheel. Don't get excited, George 2, but you might be doing some aerobatics any minute now.

" 'Prepare to corkscrew port, Jack . . . corkscrew port . . . go!'

" 'Going port.'

"I used heavy left aileron and rudder, elevators down, held the diving turn through fifteen degrees. I pulled out sharply, and turned hard to starboard halfway through a climb. George 2 responded like a PT-17—a PT-17 weighing twenty-five tons.

" 'Foxed him, Jack. He's holding off, level on the starboard quarter.'

"Protheroe then came through, 'Another bandit, skipper, four o'clock high, six hundred yards. It's an Me 210 . . .'

"Lanham broke in, 'Watch him, George, here comes number one again. Corkscrew starboard . . . go!'

"According to the navigator's log, the combat continued for another eight minutes: to me it seemed longer. After each frustrated pass, the attacker held off, content to occupy the attention of one gunner, while his partner came on in. I longed to have the heat turned down—the sweat was running down my face—but I dared not interrupt the gunners' running commentary. The sound of heavy breathing was sufficiently distracting, and I knew that it was mine.

"My wrists and forearms were reasonably strong, but I was no Charles Atlas, and George 2 wasn't feeling like a Stearman anymore. It occurred to me that these two fighter pilots were just playing games with us, biding their time until I was exhausted. Then they would rip the Lancaster to shreds. The sheet of armour plate behind me seemed pitifully small, and there was a lot of me it failed to shield. If only our Brownings had a greater range; if only I could find a layer of cloud to hide in; if only the moonlight wasn't quite so bright . . .

" 'Corkscrew port . . . go!'

"Throwing George 2 into another diving turn, I looked back through the window. There was the Messerschmitt again, turning steeply with me as the pilot tried to bring his guns to bear. I could see his helmet and his goggles, looking straight at me. Staring back at him, I felt a sudden surge of anger, and a change of mood. You're not good enough, Jerry, I thought, to win this little fight. You're a bloody awful pilot and a damned poor shot. 'Well, for Christ's sake, George, shoot that bastard down.'

" 'I think you got him,' I said. 'Where's the other one?'

" 'Falling back astern,' said Lanham. He's clearing off. Probably out of ammo or fuel.'

" 'Good shooting, George. What kept you?'

" 'Sorry, skipper. I had my sights on him all the time. I guess I just forgot to pull the trigger.'

" 'Pilot from nav. Let me know when you're back on course.'

" 'Roger.'

" 'Bomb-aimer, skip. I was ready for the buggers, but they never came in bloody range of the front bloody guns.'

"Larry was himself again. I checked the compass, and turned toward the coast of Lübeck Bay. I was thinking of the Welshman sitting in the turret with the fighter in his sights. He had fired a lot of rounds on training ranges and at air-towed drogues, but he had never fired a bullet at another human being. That was rather different, and I thought I understood why he had needed the command to open fire.

" 'Nav from pilot. Back on course. Let' all settle down.'

"I held the wheel loosely and stroked the rudder pedals with the balls of my feet. George 2 was flying head-on into the wind, and her speed across the water was a mere 176 mph. It was going to be a long ride back to Wickenby, but I believed that we would make it. At twenty minutes after one, Lanham reported in that the Peenemünde fires could still be seen.

"Forty-three minutes later, a searchlight reared ahead, pale in the moonlight but no less dangerous for that. I really hated searchlights. Over the target you just had to ignore them, but I did my best to dodge them when we were on our own. If that master beam latched on, its two slaves would quickly

Members of the 336th Fighter Squadron, 4th Fighter Group, Debden, England in their disperal hut crew room in 1944.

above: Mustang pilots of the U.S. Eighth Air Force after a bomber escort mission in 1944;
below: 357th Fighter Group Mustang pilot 'Gentleman Jim' Browning.

follow, and few aircraft, once coned, returned to base unscathed. On the way home from Hamburg two weeks earlier, we had got away with it—more by luck than judgment. I didn't want to try our luck again. 'Going ten degrees starboard for two minutes, Jim.' The beam waved toward us, like a finger feeling for a keyhole in the dark; it groped for a while and then disappeared. Back on course, George 2 began her second crossing of the cold North Sea.

"Just after three o'clock, the gee-box showed its first good signals since they faded at point A four hours ago. Cassidy plotted the position. For close on a thousand miles, by dead reckoning, a bearing on Polaris, and three or four pinpoints, he had navigated George 2 to within a mile of where she ought to be. He gave no sign of being surprised. 'Pilot from nav. We're pretty well on track. ETA base is zero-four-zero-five.'

"I liked the sound of that; now the journey could be thought of in terms of minutes, not of hours. I could begin the descent, engage the autopilot, and drink a cup of coffee. I might even struggle back to the Elsan for a piddle.

"Walker reduced the rpm. The vertical speed indicator showed a descent rate of three hundred feet a minute. If I maintained that flight path, we should arrive over Wickenby more or less at circuit height and all set for a landing, provided I remembered to lower the undercarriage. Nothing then could keep us from breakfast or from bed. Not that the gunners could relax. There could be no worse anti-climax than to get chopped by an intruder in the airfield circuit.

" 'Pilot to crew. We're below oxygen height. Smoke if you can afford it.'

"At about four o'clock the Mablethorpe searchlight stood erect, the only searchlight I was ever glad to see. I switched on the navigation lights and the IFF, and Fairbairn stood ready with the colours of the day. Ten minutes later, a beacon twinkled dead ahead. It read dit-dit-dah-dah-dit-dit, the code for Wickenby. I told Walker to turn the oxygen up to the 20,000 foot level, and pushed the R / T button A. 'Hello, Orand, this is Nemo George 2, are you receiving me, over.'

" 'Hello, Nemo George 2. This is Orand. Receiving you loud and clear, over.'

" 'Orand, George 2 approaching from the east, fifteen-hundred feet. Permission to join the circuit, over.'

"George 2, the circuit is busy. You're clear to join at 4,000 and stand by. Left-hand orbit, two aircraft at that height, over.'

" 'Shit,' said Walker. 'We're in the flipping stack.'

" 'Yeah,' snarled Myring. 'That's the bloody snag with being in the last bloody wave.'

" 'Shut up' I counseled. 'It's a lovely night for flying. Twenty-eight-fifty rpm, engineer.'

"For the next thirty minutes, George 2 orbited the beacon, gradually descending in five-hundred-foot stages at Orand's command. Later in the tour, I would learn to take some shortcuts, to make a better speed, and to arrive at Wickenby before the stack began, but in those days, a green sergeant pilot, I didn't know the score. At last the call came through: 'George 2, you'r cleared to one-thousand feet and number two to land. Runway two-seven. Queenie-Fox-Easy one-zero-one-two. Call down-wind, over.'

"I set the altimeter and began the landing drill. 'Trailing aerial in, Charles. Brake pressure, Johnny? Fuel?'

" 'Plenty of both.'

" 'Rad shutters open. Check M gear.'

"At 1,000 feet, a mile south of the field, I turned parallel with the twin lines of the runway lights, and reduced the power.

"'Orand, George 2 downwind, over.'

"'George 2, you're clear to funnels, one ahead.'

"'Wheels down, Johnny.'

"The undercarriage lamps shown red as the up-locks disengaged, and the nose dropped a fraction as the airflow hit the wheels. The locks engaged with a jolt, and the lamps turned to green. 'Flap fifteen. Booster pumps on.'

"When the last set of lights at the runway's downwind end were level with the port wingtip, I brought the airspeed back to 140 and turned toward the field. Halfway through the turn, the funnel lights on the port quarter beckoned like the gates of home. As the nose swung into line, I inched the throttles back and let gravity do its stuff. 'George 2, funnels, over.'

"'George 2, you're clear to land. Wind is eighteen degrees from your right at ten knots, over.'

"'Half flap, Johnny. Pitch fully fine.'

"I held the nose down to counteract the lift and steered a mite to starboard to compensate for drift. The lights of the runway seemed to widen at the threshold and to taper in the distance up ahead. 'Full flap. Stand by for landing.'

"As George 2 crabbed across the threshold, Walker held the throttles back against the stops. I kicked the nose straight and pulled the wheel into my lap. The tyres squealed on the tarmac at 04:49.

"The record showed that 560 aircraft reached the target that night and dropped 1,800 tons of bombs, eighty-five percent of which were high explosive. When the truth was revealed about the Baltic base, it was said that Hydra set the V-weapon programme back by several months and reduced the scale of the eventual attack. Certainly, no flying bombs fell on England until June 1944, and no rockets until the following September. General Dwight Eisenhower wrote later: "If the Germans had succeeded in perfecting and using these new weapons six months earlier, our invasion of Europe would have proved exceedingly difficult, perhaps impossible."

"One-hundred and eighty Germans, mostly scientists and technicians, died in the attack and General Jeschonnek, the Luftwaffe Chief of Staff, committed suicide the next day. Sadly, the first Newhaven markers went down on the camp where the slave workers were sleeping, and over five-hundred unhappy lives were lost.

"The Nachtjagd, it transpired, had deployed a new device: twin upward-firing cannon mounted behind the cockpit of the Me 110 and fired by the pilot with the aid of a reflector sight, enabled the fighter crew to attack the bomber's blind spot underneath the fuselage. This deadly piece of weaponry, known as *Schräge Musik*, was believed to have inflicted some of the losses suffered on the night: twenty-three Lancasters, fifteen Halifaxes, two Stirlings, and one of the Mosquitos.

"For the future, an airborne MC or master bomber, would control all major raids, and the innovative tactics for 'recentering' ground markers were to be retained. Peenemünde would be the target for three U.S. Eighth Air Force missions in July and August 1944, but Hydra was, and would remain, the RAF's only full-scale precision attack in the last two years of the war.

"Knowing nothing of these matters, we drank our cocoa with a tot of rum and attended the debriefing. The crew were in good spirits: we had hit a vital target, dodged the searchlights and the flak, outflown one Messerschmitt and destroyed another—well, possibly destroyed. It seemed a shame to remind them, as we ate our eggs and bacon, that we hadn't yet completed one third of our tour."